WIRING SIMPLIFIED

Based on the 2014 National Electrical Code®

44th edition

Frederic P. Hartwell

W. Creighton Schwan

H. P. Richter

PARK PUBLISHING, INC.
Minneapolis, Minnesota
New Richmond, Wisconsin

PARK PUBLISHING, INC.
511 Wisconsin Drive
New Richmond, Wisconsin 54017

Printed in the United States of America

ISBN: 978-09792945-5-6

Warnings and Disclaimer
This book provides useful general instruction, but we cannot anticipate all possible working conditions or the characteristics of your materials and tools. For your safety, and the safety of others, you must use caution, care, and good judgment when following the procedures described in this book. You must adapt the instructions to your particular situation. Consider your own skill level and follow the instructions and safety precautions and warnings associated with the various tools and materials shown. Electricity and certain tools can be dangerous and can cause injury.

Information has been obtained from sources believed to be reliable, but its accuracy and completeness are not guaranteed. The publisher and author disclaim liability, whether for personal injury or property damage, arising out of typographical error or the use or misuse of this publication regardless of the fault or negligence of the publisher or author. The instructions in this book conform to the 2014 *National Electrical Code*®. Consult your local authorities for information on building permits, codes, and other laws as they apply to your project.

Executive editor: Steven F. Wolfe
Editor: Marly Cornell
Indexer: Galen Schroeder
Layout: The Roberts Group

8 7 6 5 4 3 2 1

Contents

Tables

Preface to the 32nd Edition

This book has been written for people who want to learn how to install electrical wiring so that the finished job will be both practical and safe. The installation will comply with the *National Electrical Code®*. Then the finished job will be acceptable to electrical inspectors, power suppliers, and others having jurisdiction in the matter.

Electrical wiring cannot be learned by skimming through this or any other book for fifteen minutes; neither should the book be considered a "Quick Answer" reference. Nevertheless, careful study of this book should enable you to wire a house or a farm so that it will be acceptable to everyone concerned. However, before doing any wiring, learn how to do the job correctly.

The author hopes this book will also be of considerable value in planning a wiring job, to enable you to write sensible specifications that will lead to your securing maximum usefulness from electric power—now, and ten years later. Careful planning will avoid changes that usually cost several times as much as when included in the original plans.

Throughout this book I have emphasized the reasons why things are done in a particular way. This will help you understand not only the exact problems discussed, but will also help you solve other problems as they arise in actual wiring of all kinds.

The first edition of *Wiring Simplified* appeared in 1932. This 32nd edition will be prepared by W. Creighton Schwan, who is especially well qualified to do so. I am very grateful to him for undertaking this project.

—H. P. Richter

Preface to the 39th Edition

The ability of H. P. Richter to reduce a complex subject to simple, easy to understand terms is rare. Matching that ability is improbable, but it is a challenge I seriously accept. This 39th edition reflects revisions made necessary by the many changes in the 1999 *National Electrical Code®* and by materials and methods not covered in previous editions.

—W. Creighton Schwan

Preface to the 44th Edition

It is a privilege to begin my fifth *NEC* cycle as the author, having taken over the project for the 2002 *Code* cycle. W. Creighton Schwan did an outstanding job maintaining the continuity of this work from the 32nd edition, which reflected the developments in the 1975 *NEC*, through the 39th edition based on the 1999 *NEC*. My continuing hope is to maintain what my predecessors managed so admirably in keeping an increasingly complex field accessible to a broad audience. This 44th edition reflects the changes in the 2014 *NEC*. The 2014 *NEC* resulted from Code-Making Panel actions on 3,745 proposals from the public, which after publication provoked 1,625 public comments. Every proposal and every public comment required individual attention, including a vote and a written technical explanation in those many instances where the suggestion was not accepted in full. The result is a continuously evolving document that fully reflects the rapid pace of changes in our technological landscape, carefully sifted through a consensus review process that is second to none in inclusiveness and technical accuracy. This edition of *Wiring Simplified* fully tracks the *NEC* changes that are within its purview.

This edition, as always, responds to many helpful suggestions by readers, and further comments and suggestions for improvements are greatly appreciated. For contact information, see page xii.

—Frederic P. Hartwell

Introduction and Guide

Park Publishing, Inc. specializes in wiring books that offer easy-to-understand instructions along with valuable guidance from *National Electrical Code*° experts. Throughout its history of more than seventy-five years of continuous publication, *Wiring Simplified* has been revised to conform to each new *National Electrical Code*. Based on the 2014 *NEC*, this 44th edition of *Wiring Simplified* has been fully updated by recognized *NEC* expert and licensed master electrician Frederic P. Hartwell. The next edition of *Wiring Simplified* will be based on the 2017 *NEC*.

INTENDED AUDIENCE

Wiring Simplified has long been used with success by a variety of readers—

- The person wiring a new home or updating existing wiring in an older home will learn whether the proposed work conflicts or accords with or even exceeds *NEC* requirements. Step-by-step directions are given for completing a safe, convenient, and practical installation.

- The homeowner employing a professional will find this book invaluable in discussing the job with the contractor. Studying *Wiring Simplified* will give the homeowner insight into the problems the contractor will need to solve and will provide the basis for clear communication.

- The person wanting to troubleshoot and make electrical repairs will find instructions in a special chapter.

- Students and teachers using the book in a variety of educational settings appreciate the range and depth of subject matter. *Wiring Simplified* is a dynamic text that gives students a sense of being on the job with a professional.

GUIDE TO USING THIS BOOK

The nineteen chapters in *Wiring Simplified* are divided into four parts that cover the various phases of a wiring job—from understanding basic vocabulary and principles, to planning, installation, and finishing. Each part provides the necessary foundation for the parts that follow.

The best way to use this book is to read it through from start to finish to gain

National Electrical Code° and *NEC*° are registered trademarks of the National Fire Protection Association, Inc., Quincy, MA 02169.

an overall understanding of what is involved in electrical wiring. This step is vital to fully grasp the elements of a complicated trade. Many questions from readers turn out to address topics that are fully covered in the book, but the reader may have neglected to read through the book first.

Use these tools that will help you find and understand the information in the book.

■ **TABLE OF CONTENTS**—gives the page number for each chapter and provides an overview of the wide range of material covered in the book.

■ **INDEX**—your key to locating specific topics quickly and easily.

■ **GLOSSARY**—a quick guide to the electrical terminology that is used throughout the book, to assist readers with unfamiliar terms. It is crucial that no term in this book be used in a way that is incompatible with its usage in the *National Electrical Code*. Electrical terms often differ from everyday usage. The terms in this book are no exception. To take just a few examples: a "bus" is not a vehicle; a "fish tape" is not tape in conventional terms; a "bond" is not a financial instrument, and a "building" for these purposes may actually be one among many contiguous structures combined into something much larger that in normal conversation would be collectively referred to as a single building. The glossary begins on page 231, and should be consulted as necessary, or even reviewed in advance in its entirety.

The four parts of the book are—

1. GETTING STARTED—THE BASICS, CHAPTERS 1–3
2. WIRES, CIRCUITS, AND GROUNDING, CHAPTERS 4–7
3. INSTALLING SERVICE EQUIPMENT AND WIRING, CHAPTERS 8–15
4. SPECIAL WIRING SITUATIONS AND PROJECTS, CHAPTERS 16–19

PART 1. GETTING STARTED—THE BASICS provides basic background necessary to understand the safety, practical, and legal considerations involved in wiring decisions and practices. This is essential reading before undertaking any electrical work. The standards and codes that help ensure safety are explained, and important safety precautions are offered. Practical goals to consider in the planning of any wiring job are described in plain language. A bare-bones outline of how electrical power is measured and delivered to the home provides basic vocabulary for understanding issues of usage, cost, and conservation of electrical power.

PART 2. WIRES, CIRCUITS, AND GROUNDING explains how to choose the right wire types and sizes, electrical devices, and tools you'll need for a particular job. Proper overcurrent protection and the importance and specifics of grounding are emphasized here and throughout the book. New developments on arc-fault circuit interrupters are covered. Regulations are given for feeders and the panels they supply. A simple way is described for using wiring diagrams to design installations of any size.

PART 3. INSTALLING SERVICE EQUIPMENT AND WIRING When you're ready to start your installation, *Wiring Simplified* fits in your toolbox so you can take it right to where you're working for ready reference. All commonly encountered wiring methods are covered. The service entrance chapter describes required workspace around panels and explains how to establish the connection point between the power supplier's wires and your electrical installation. Detailed instructions are provided for selecting and installing boxes and switches in your circuits; for making the wiring connections at the boxes, switches, and receptacles; for running cable or raceways; and for testing the completed installation. You will also find information on how to wire detached garages and outbuildings. Modernizing an outdated wiring system is covered in a separate chapter, with many expert hints for making the job easier. Wiring requirements for a variety of specific household appliances are listed. The details of finishing an electrical installation wrap up this professional approach to getting the job done.

PART 4. SPECIAL WIRING SITUATIONS AND PROJECTS begins with a discussion of the types and wiring requirements of stand-alone motors commonly used in residences and on farms, with special coverage of new regulations for motor disconnects. The chapter on farm wiring covers specialized topics such as constructing the meter pole, wiring animal buildings, and rural safety issues. In addition to telephone and doorbell wiring, the chapter on low-voltage wiring includes general material on computer network wiring. The last chapter, on troubleshooting and repairs, is a guide to diagnosing and remedying common problems such as blown fuses or tripped breakers, broken switches and receptacles, and lamps and doorbells that don't work—information that can be used by people who never intend to undertake an electrical installation but want to be able to maintain the electrical equipment they depend on each day.

Wiring is an exacting practice and deserves your best effort. Consult your local electrical inspector before you start work, and learn which codes are in effect in your locality. We hope this book will be useful in addressing your primary concerns while doing electrical work safely (including disconnecting live circuits, standing on dry surfaces, and using safety glasses), and in creating installations that are safe for ongoing use.

Your questions, comments, or suggestions about this book are welcome.

Please send them to—

Park Publishing, Inc. Phone 800-841-0383
511 Wisconsin Drive Fax 715-246-4366
New Richmond, WI 54017 Email ParkPublishing@nrmsinc.com

Chapter 1
STANDARDS, CODES, AND SAFETY

WE DEPEND ON ELECTRIC POWER so extensively in our homes and businesses, on our farms, and in industry that it would be difficult to list all its uses. Electric power serves in countless ways while normally presenting little danger, and we often take it for granted. Yet when not used properly, electric power can cause fire, destroy property, seriously injure people, and even cause death.

STANDARDS
Electric power is safe for us to use only when it is under control. We control it by using wiring and electrical equipment that is of dependable quality. Equally important, the wiring and electrical equipment must be properly installed and maintained.

Electrical parts and devices are manufactured according to specific safety standards. They should be installed using methods that are uniform throughout all the states—the methods proved by experience to be both practical and safe.

Product listings Reputable retailers and manufacturers sell only merchandise that is listed by a qualified electrical testing laboratory. The oldest and most commonly recognized testing organization is Underwriters Laboratories (UL). When a manufacturer submits its product to such an organization, the product is investigated and subjected to performance tests. If it meets the safety standards, it is then "listed by (name of laboratory)." Many people say "approved by," but that is not correct terminology. To determine that the product continues to meet requirements in the product standards, trained field representatives periodically visit the factory where the product is made to audit production controls. Additionally, the testing laboratory regularly tests factory and store samples. If these samples meet the requirements, the product continues to be listed and will bear a listing mark.

Some items (wire, large switches, fixtures, conduit) have a listing mark similar to those shown in Fig. 1–1 on each coil or piece. Cord sets (extension cords) have labels as shown in Fig. 1–2. Devices like toggle switches carry the name of Underwriters Laboratories Inc. (may be abbreviated) and the word LISTED molded or stamped on each piece. Still other items such as receptacles, sockets, outlet boxes, and toasters have no label; but the nameplate, tag, or the item itself bears the listing mark of Fig. 1–3. Each piece is marked in some way so that the inspector or purchaser can identify the manufacturer by referring to a directory listing manufacturers that have proved their product meets the applicable requirements. The listing mark on the product indicates compliance. In addition to UL, other certification laboratories include ETL Testing Laboratories, Inc. (now part of Intertek Testing Services), Factory Mutual (FM), and the Canadian Standards Association (CSA); however, UL is the most widely used.

"Listed" usage requirements The term "listed" indicates the merchandise is suitable and safe *if used for the purpose and under the conditions* for which it was designed. Electrical inspectors will turn down listed merchandise if improperly used. For example, the inspector will refuse to accept listed armored cable used in a barn if the barn is a corrosive location, because armored cable is for use only

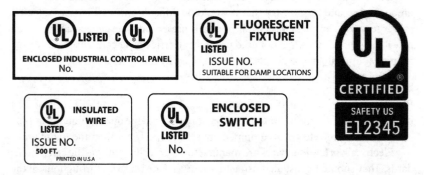

Fig. 1–1 Examples of labels applied on merchandise listed by Underwriters Laboratories Inc. The label to the right is an example of the next generation of UL labels, using the word "certified" and modular elements. The term is more easily understood in international commerce. Its use is not mandatory and all existing marks are and will remain valid.

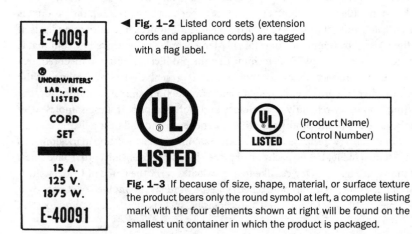

◀ **Fig. 1–2** Listed cord sets (extension cords and appliance cords) are tagged with a flag label.

Fig. 1–3 If because of size, shape, material, or surface texture the product bears only the round symbol at left, a complete listing mark with the four elements shown at right will be found on the smallest unit container in which the product is packaged.

in permanently dry locations. Listed lamp cord will be turned down if used for permanent wiring because such cord is listed only for use on portable equipment. Listing by UL does not mean that similar pieces of merchandise are the same quality. It only indicates that the pieces meet UL safety requirements. One piece may far surpass those requirements while another barely meets them. Of two brands of toggle switches, one may average 25,000 ons and offs at full load before breaking down, and the other only 6,000. Among listed brands, base your choice on desired features, quality, and price, as you would for any other merchandise.

CODES, PERMITS, AND LICENSES

Listed electrical parts of high quality that are carelessly or improperly installed might still present a risk of both shock and fire, so rules have been developed regulating the installation of wiring and electrical devices.

National Electrical Code®   For a safe installation, listed devices must be installed as required by the *National Electrical Code* (abbreviated *NEC*). The *NEC* is simply a set of rules specifying the installation and wiring methods that over a period of many years have been found to be safe and sensible. The *NEC* permits installations to be made in several different ways, but all wiring must be done in one of the ways described in the *NEC*. A revised edition of the *National Electrical Code* is published every three years, with the 2014 edition as the most current. Every reader is urged to obtain a copy of the *NEC* and to use it as a reference. It contains detailed regulations and extensive tables. Because the *NEC* is more than 800 pages long, only its most important general points can be included in *Wiring Simplified*.

In this book, all reference to *NEC* section numbers will be to sections in the 2014 *NEC*. A reference, for example, to "90.1" indicates Section 1 of Article 90. The 2014 *NEC* contains a large number of changes from the 2011 edition, including: (a) renumbering of some sections without a change in text; (b) renumbering of some sections with changes in the text, sometimes of minor but often of major importance; and (c) new section numbers covering material not in the previous edition. Therefore, a 2014 section number mentioned in this book may or may not be about the topic covered by the same section number in an earlier or later *NEC*. Highlighted text on *NEC* pages indicates material that differs from the previous edition. The marginal bullets indicate deleted material, but only if an entire paragraph or more was deleted.

All methods described in this book are in strict accordance with the latest *National Electrical Code* as interpreted by the author. Every effort has been devoted to making all statements in the book correct, but the final authority on the *NEC* is your local inspector. This book covers only the wiring of houses

> National Electrical Code® and NEC® are registered trademarks of the National Fire Protection Association, Inc., Quincy, Massachusetts. Every reader is urged to study the *NEC*. You can obtain a copy by sending $89.50 plus $8.95 shipping and handling to:
>
> NFPA Fulfillment Center, 11 Tracy Dr., Avon, MA 02322-9908 or by calling 1-800-344-3555. It is also carried in many large book stores. Ask for *NEC* 2014.

and farm buildings. Any statement that some particular thing is "always" required by the *NEC* means "always so far as the type of wiring described in this book is concerned."

Study of the *NEC* is necessary and helpful, but the *NEC* alone will not teach you how to wire buildings. Read *NEC* 90.1 for the exact wording of its purpose, which is essentially the practical safeguarding of persons and property from hazards arising from the use of electricity. These hazards are fire and electric shock. The *NEC* contains provisions considered necessary for safety. Compliance with the *NEC* together with proper maintenance will result in an installation essentially free from hazard but not necessarily efficient, convenient, or adequate for good service or future expansion. The *NEC* is not intended as a specification for design, nor is it intended as an instruction manual for the inexperienced. A lay reader picking up a copy of the *NEC* without specialized training will likely be hopelessly confused. Electricians generally spend at least four years in training, including extensive work with *Code* instructors, before they work in the field without supervision.

Legal aspects Neither the *NEC* nor the product standards as enforced by the testing laboratories have the force of law. However, most states (and/or municipalities if so authorized by the state legislature) pass laws requiring that all wiring must be in accordance with the *National Electrical Code*. Usually power suppliers will not furnish power to buildings that have not been properly wired, and insurance companies may refuse to issue policies on buildings not properly wired. In addition, faulty work may complicate insurance reimbursements in the event of a loss. You have no choice except to follow the law, and in doing so conscientiously you will automatically produce a safe installation.

Local codes The *National Electrical Code* is sometimes supplemented by local codes or ordinances, which seldom differ from the national *Code* in general terms but which frequently amend specific provisions. For example, armored cable wiring is one method permitted by the *NEC* but sometimes prohibited by local codes.

Local permits In many places it is necessary to get a permit from city, county, or state authorities before a wiring job can be started. The fees charged for permits generally are used to pay the expenses of electrical inspectors, whose work leads to safe, properly installed jobs. Power suppliers usually will not furnish power until an inspection certificate has been turned in.

Local licenses Many areas have laws stipulating that no one may engage in the business of electrical wiring without being licensed. Does that mean that you cannot do electrical work on your own premises without being licensed? In some localities the law is interpreted that way, but not in others. Consult your power supplier or local authorities.

Remember that if a permit is required, you must get one before proceeding with your work. Before applying for a permit, be sure you understand all problems in connection with your job so that your wiring will meet national and local code requirements. If it does not, the inspector must turn down your job until any errors have been corrected.

SAFETY PRACTICES

Electrical systems in buildings must be installed in such a way that they present the least possible hazard to the occupants or the property. Any unique conditions and uses, such as encountered on farms, must be carefully evaluated for wiring requirements that promote safety for people, animals, equipment, and buildings. In addition, the work itself must be performed safely for the protection of the installer. Unless you feel confident that you can satisfy these safety objectives, it would be best that you use this book as a guide to understanding your wiring system and leave the actual installation to a professional. At the very least, this book will allow you to have a more intelligent conversation with your electrical contractor, and get more for your money.

Protect your eyes   For your personal safety whenever you are doing any electrical work, protect your eyes at all times by wearing safety glasses. Bits of copper can fly out when cutting wire, and hot solder, flux, and plaster dust have a way of making right for the eyes. If an arc is inadvertently started when you must test on an energized circuit, molten metal can be thrown out too quickly for you to escape it.

Be aware of electrical fire hazards   Short circuits, overheating, and the arcing or active flow of electricity through an ionized path in the air or other gas (and usually where it doesn't belong)—these are hazards that can result from faulty wiring, and cause fire. Electrical arcs, for example, typically run at temperatures considerably above that on the surface of the sun. Be careful to use wires with the proper ampacity (explained in Chapter 4) and check for wires or cords with damaged insulation. In Chapter 7 you will find a thorough discussion of the essentials of proper grounding. Protecting your electrical installation through the use of overcurrent devices—fuses and circuit breakers—is the focus of Chapter 5. Limiting vulnerability to lightning damage is a special concern for rural residents and is discussed in Chapter 17, "Farm Wiring." Safe wiring practices for fire prevention are emphasized throughout the book.

Take advantage of hazard warning detectors   Faulty wiring is just one of many possible sources of fire or other hazards in a home. Installing appropriate alarm equipment is a relatively small investment that can save lives. Regularly check to see that your warning alarms are functioning. Get in the habit of replacing the batteries in a smoke detector or CO (carbon monoxide) detector every fall at the same time as you reset your clocks to standard time. Remember that modern designs of even permanently wired detectors usually depend on battery backup.

   Smoke detectors   Self-contained smoke detectors that sound an alarm when visible or invisible combustion products are sensed have been credited with saving many lives by alerting occupants early enough for them to escape from a fire. Smoke detectors in homes are required by local building codes. There are three types of detectors. *Ionization alarms* are especially effective at detecting fast-flaming fires that quickly consume combustible materials such as burning paper or grease. *Photoelectric alarms* are more effective at detecting slow, smoldering fires such as a cigarette burning in a couch, which may take hours before bursting into flame. For your maximum protection, install both types or select the third kind— a *dual detector* that combines both features. Building codes in each community specify the required locations but generally not the type of alarm. In general, never place smoke detectors where they will be subject to nuisance alarms. For example, an ionization detector placed near a bathroom door will likely go into alarm when the door opens following a shower. Some local codes require a smoke detector in the hallway outside bedrooms or above the stairway leading to bedrooms on an upper floor, while other localities require a detector in each bedroom, the kitchen, and one on each floor of the dwelling.

Battery-operated, plug-in, and direct-wired smoke detectors are available. Battery-powered models may be acceptable in an existing dwelling, but in new construction most building codes require connection to the electrical source, with a battery backup and all alarm units interconnected by wires so that if one responds, they all respond. The required location is on the ceiling, or on the wall not more than 12 inches from the ceiling. Don't place the detector in the path of supply air ventilation that would move clean air past the detector faster than smoke could reach it. To install a direct-wired detector, choose a circuit with often-used lights on it, such as the bathroom light, to assure the circuit could not be off without being noticed. Wire the detector directly across the two circuit wires, unswitched.

Carbon monoxide (CO) detectors   Many local ordinances also require permanently wired carbon monoxide detectors with battery backup to alert occupants to the presence of this invisible, odorless, deadly gas. Check with your building official or fire marshal for specific requirements. CO detectors are highly recommended even if not required by local codes. Battery-only styles are available. CO detectors can usually be found in stores near the smoke detectors.

Prevent electric shock   Safe installation practices are emphasized throughout this book. Shock—the accidental flow of electric current through the body—is the principal hazard that can result from unsafe practices. Observing the following cautions will help you avoid injury.

Disconnect the circuit   Whenever possible, work should be done only on circuits that have been de-energized (disconnected from the source of power). If the disconnecting means is out of sight from the work location, you must take positive steps, such as taping the switch to the OFF position or removing the fuse, to make sure someone else cannot inadvertently energize the circuit while you are working on it. Some testing can be done only on energized circuits. Be extremely careful when testing that you do not contact live parts. As explained on page 65, receiving a shock is more dangerous when the victim is standing in water or a damp location than on a dry surface. For your protection, stand on dry boards or a rubber mat, and be sure that neither you nor the floor is wet. Remember that electricity and water are a dangerous combination.

Avoid shocks from small appliances   Even when a building has been wired in strict accordance with the *NEC*, shocks can still occur from the use or misuse of appliances. Fatal shocks have occurred in bathtubs or showers when an appliance such as a radio or hair dryer has accidentally fallen into the water. In that situation, even an appliance that is not defective can cause a fatality. Touching a defective radio or heater or other appliance while in a tub or shower can also be fatal. A child might chew on the cord of a lamp or appliance; in doing so, if the child punctures the insulation of the cord (or if the cord is defective) while he or she is also touching a grounded object such as a radiator or plumbing, a fatal shock can result. A person using a defective appliance while touching a grounded part of plumbing can easily receive a shock. (See page 74 for a discussion of GFCI protection.)

A toaster can present extreme danger if the toast sticks rather than popping up when it is done, because when that happens people sometimes use forks to release their toast. Don't do it without first unplugging the toaster. The live heating elements are exposed in the toaster. If your fork touches a heating element and you are touching a grounded object (such as a water faucet or a sink) at the same time, you could receive a dangerous or even fatal shock. Unplug the toaster before reaching inside it for any purpose, or use wooden tongs to extract the toast.

Know how to help shock victims safely   If you discover a shock victim, the most important point is to be careful not to become a second victim yourself. This book cannot possibly give you all the needed information, but it can give you a few hints.

We'll first discuss shocks caused by the ordinary 120/240-volt wiring in a house or a farm building. If you find a victim and it appears that the shock was caused by an appliance or similar equipment, disconnect that equipment. If it has a wall switch, turn it off. If it doesn't, pull the plug out of its receptacle; do not touch the cord because it may be defective, but rather grasp the plug and pull it out. Do not touch the appliance itself.

In cases where the source of the shock is not apparent, hurry and open the main switch of the building, or throw the main circuit breaker to OFF if your building is equipped with breakers. Then turn your attention to the victim. If the victim is unconscious or appears not to be breathing, phone 911 or the emergency service in your area for assistance. If you know CPR (cardiopulmonary resuscitation), start using it while you are waiting for medical help. (You'll be better prepared for emergencies if you have taken a class in CPR. Contact your local Red Cross chapter for information.)

If the shock has been caused by a high-voltage line that has been knocked down by a storm or accident, the situation is quite different and many times more dangerous. In such a case, phone 911, and the power company if possible. If the high voltage wire is touching the victim, do not touch him or her; if you do, you could become the second victim by trying to help somebody who quite possibly has received a fatal shock. Theoretically you can use a completely dry wooden pole at least four feet long to push the fallen wire off the victim, but such accidents seldom occur near your home, and you might not have such equipment with you. You might be able to rig a sling or rope (only if absolutely dry) to separate the wire from the victim. Time will be a critical factor. If at all possible, depend on experts from your power company to perform the rescue.

Chapter 2
PLANNING YOUR ELECTRICAL INSTALLATION

THE PREVIOUS CHAPTER ON STANDARDS and codes discussed the importance of following the *National Electrical Code* (*NEC*) strictly in order to produce a safe wiring installation. This chapter outlines how careful planning will help you make it a convenient, efficient, and practical installation. Here you will find the characteristics of an adequate house wiring installation, the principles of good lighting, and *NEC* requirements for locating lights, switches, and receptacles.

CONSIDER PRESENT AND FUTURE NEEDS
Plan your electrical installation so you will still be pleased with it in the years ahead. Do not skimp on the original installation. Adding outlets, receptacles, switches, fixtures, and circuits later usually costs several times more than it would to include them in the original job. Look ahead to the future and try to allow for electrical appliances you do not as yet own.

By installing large wires and extra circuits now, you will have adequate wiring in the future. You will be making a good investment.

Install large service entrance   All the power you use comes into the building through the service entrance wires and related equipment. Start your planning with a service entrance that will adequately handle present and future needs. Keep in mind that small service wires will not carry a large load satisfactorily. Chapter 8 presents a full discussion of the service entrance installation.

Benefit from larger circuit wires   For house wiring, the minimum circuit wire size permitted by the *NEC* is 14 AWG, protected by a 15-amp fuse or circuit breaker, except for door chimes and other low-voltage wiring described in Chapter 18. There is a trend toward using 12 AWG wire, with 20-amp protection; and in a few localities it is required as a minimum. The larger wire means brighter lights, less power wasted in heating of wires, and less frequent blowing of fuses or tripping of breakers. The use of 12 AWG wire in place of 14 AWG adds very little to the cost of the installation and will prove a good investment. *NEC* requirements for various types of circuits are discussed starting on "Calculating Branch Circuit Needs" on page 47. Special circuits requiring 12 AWG wire are explained under "Special small-appliance circuits" on pages 48–49.

Note that the *NEC* uses the more technically correct term "AWG" (as in 12 AWG, standing for "American Wire Gauge") instead of the old term "No." (for "Number" as in No. 12), and this book reflects that practice. Wire gauge numbers (now expressed as "___ AWG") apply for wiring up to 4/0 AWG, which covers almost all wiring within the scope of this book. Wires above that size are denominated by their actual cross-sectional area, as in 500,000 circular mils (written 500 kcmil), the usual size for a 400-amp service conductor.

LOCATION OF LIGHTING, SWITCHES, AND RECEPTACLE OUTLETS

An adequately wired home is well lighted. The lighting will be from floor and table lamps in some rooms and from permanently installed fixtures in others. Lighting fixtures do not need to be expensive or elaborate, but you must have outlets for fixtures located to provide sufficient light where needed. In addition, plenty of switches and receptacle outlets are essential for convenience and safety.

Height from floor for switches and receptacles   There is no *NEC* requirement setting how high switches and receptacles must be from the floor, although switches used as disconnecting means for electrical equipment generally must not be higher than 6 feet 7 inches above the floor or working platform. The typical range is 48 to 52 inches for switches and 12 to 18 inches for general-use receptacles. (People with physical disabilities may require switches to be lower and receptacles higher than usual.) Kitchen receptacles are located in the backsplash above the counters. In the laundry room and the workshop, a convenient height for plugging in washer, iron, and tools is 36 to 48 inches above the floor. In new construction, make the heights uniform throughout. In a remodel or addition, match the existing heights.

NEC requirements for lighting and switches   *NEC* 210.70(A) requires the following areas to have some lighting controlled by a wall switch: every habitable room of a house; bathrooms; hallways; stairways (switches at both ends if comprised of at least six risers); attached garages, and detached garages if wired; attics, underfloor spaces, utility rooms, and basements used for storage or containing air-conditioning, heating, or other equipment requiring servicing. In kitchens and bathrooms the switch must control permanently installed lighting fixtures. In other rooms the switch may control one or more receptacle outlets into which floor or table lamps can be plugged. Each entrance into the house must have an outdoor light controlled by a switch inside the house.

Regardless of *NEC* requirements, an adequately wired house will have wall switches located so you can come in through any entrance and move throughout the house from floor to floor without ever being in darkness or leaving lights turned on behind you. A wall switch (shown in Fig. 2–1) costs more than a pull chain, but the wall switch is far more convenient to use, and it provides accident insurance

Fig. 2-1 Typical toggle switch. The "plaster ears" on the ends of the strap are a great convenience in mounting the switch. *(Pass & Seymour/Legrand)*

against stumbling while looking for a pull chain in the dark. Installing a pair of three-way switches so a light can be turned on or off from two different places (the bottom and top of a staircase for example, as now often required) costs only a few dollars more than a single switch, and is well worth the difference in terms of safety and convenience. Although the *NEC* does not specify switch locations, providing convenient switches for lighting also helps to conserve energy since we are less likely to leave lights on if it is easy to turn them off when they are not needed.

NEC requirements for receptacles *NEC* 210.52(A) requires receptacle outlets in each kitchen, family room, dining room, living room, parlor, library, den, sun room, bedroom, recreation room, or similar rooms (or areas, as in the case of an open floor plan) of dwelling units to be installed so that no point along the floor line in any wall space (including spaces behind door swings) is more than 6 feet measured horizontally from an outlet in that space, including any wall space 2 feet or more in width and the wall space occupied by fixed glass panels in exterior walls. The wall space provided by fixed room dividers, such as freestanding counters or railings, is to be included in the 6-foot measurement. Hallways 10 feet or more in length must have at least one receptacle outlet. Large foyers (over 60 ft²) generally require at least one receptacle on any contiguous wall space over 3 ft wide. A receptacle controlled by a snap switch does not meet these spacing requirements, so the normal approach is to split the receptacle (as covered on page 116). The always-on receptacle meets the spacing rule, and its switch-controlled sister meets the lighting rule. Note also the special rules on tamper resistance and weather resistance covered in Chapter 10.

Basements are required to have at least one receptacle outlet in addition to any outlets for laundry equipment. Attached garages—and detached garages having electric power—are also required to have at least one receptacle outlet. Every single-family house (and both units of a duplex if both are at grade) is required by *NEC* 210.52(E) to have at least two outdoor receptacles—one on the front of the dwelling and one on the back—for yard tools, appliances, holiday lighting, etc. They must be no more than 6 feet 6 inches above grade level. See page 74 for locations where GFCI protection for personnel is required. Multifamily (three or more units) row housing units need only one outlet at either the front or the back.

An adequately wired house has enough receptacles (shown in Fig. 2-2) so that you will never need an extension cord for a lamp, clock, or similar equipment. Extension cords are useful as temporary devices, but they should never be used to

Fig. 2-2 A duplex receptacle, rated at 20 amps, 125 volts. The more common 15-amp version is identical in configuration except it does not have the horizontal slots extending to the right. The duplex receptacle has two terminal screws on each side, plus one green grounding screw (in this case at the upper right and not visible in this photo). The terminal screws at the left (ungrounded terminals) are finished with a brass or copper color; the screw terminals to the right (not visible) are for the grounded circuit conductor, and have a white or silvery white finish. *(Pass & Seymour/Legrand)*

permanently carry power from a receptacle to lamps or other equipment. If there are circumstances where you cannot avoid using an extension cord, don't invite a fire by running the cord under a rug or carpet. Never run one unattended across open floor space—stepping on it could result in damage and fire, and tripping on it could lead to injury. Never tack or staple a cord to a wall. It is best to reserve extension cords only for temporary use.

Fixtures and receptacles room by room   The following list details both the required and the recommended lighting, switches, and outlets to be installed in each room of the house. Garages and outbuildings are discussed at the end of the chapter.

Living room   Suit yourself about whether to have a fixture in the center of the ceiling or on a wall. Some homes depend entirely on floor and table lamps for living room light. If this is the case, it will be convenient to have one or two wall switches control all the receptacles into which the lamps are plugged so they don't have to be turned on and off individually. You might choose to have just the bottom or top halves of several duplex receptacles in a room controlled by a switch (See pages 116–117).

No matter how small the room, there should be a minimum of five receptacle outlets for uses such as TV, stereo, radio, and lamps. Locate one receptacle where it will always be completely accessible for the vacuum cleaner regardless of the furniture arrangement. Many living rooms will be large enough to require more than five receptacles based on the "6 foot" rule. Regardless of room size, consider the probable arrangement of the furniture, and place receptacles so they will not be difficult to reach. This may require an extra receptacle or two, but will be worth it.

Dining room   One ceiling outlet for a fixture should be provided; it must be controlled by a wall switch. Locate this fixture above the center of the dining room table rather than the center of the room. The "6 foot" rule usually will require at least four receptacle outlets. One receptacle should be always accessible from the dining room table for use with items like electric coffee makers, electric woks, and electric knives.

Kitchen   Abundant wiring is required in the kitchen. First, it should be well lighted because it is often the center of activity. A ceiling fixture controlled by a wall switch must be provided for general lighting. (The *NEC* in its 2002 edition adopted the international term "luminaire" to replace former references to fixtures, but until that term is in more general use this book will continue the traditional usage.) Another light above the sink is necessary so a person does not have to work in his or her own shadow. Consider other kitchen work areas that may need specific lighting.

All fixtures with exposed conductive parts must be grounded. If there is no equipment ground available, it must be provided, generally through rewiring the circuit. On existing wiring, only in circumstances where the wiring method provides no equipment ground, the *NEC* does allow GFCI protection in lieu of an equipment grounding connection. If you don't meet at least one of these provisions, you can only use fixtures with porcelain or plastic bodies, a category offering little in the way of variety. These are safety measures. An ungrounded metal fixture or metal pull chain could become energized through a fault in the fixture and make a serious shock possible.

The *NEC* requires two special circuits to handle only receptacle outlets for small appliances (including refrigeration equipment) in kitchen, pantry, dining room, and breakfast room (and similar areas). Both of these required small-appliance branch circuits must appear at the kitchen counter outlets (meaning all counter outlets taken collectively, not necessarily both circuits present at each outlet), because that is where the majority of appliances are connected. Install one outlet for the refrigerator, and place the others 6 to 10 inches above the counter in the backsplash. A counter space 12 inches wide requires one receptacle, and others are required so that no point along the counter space is more than 24 inches from a receptacle outlet. Spaces separated by a sink or range top are considered as separate spaces. Because the lack of a backsplash makes receptacle placement at counter peninsulas and islands difficult, the spacing requirement is relaxed to require only one receptacle outlet per counter. Be aware that where receptacles are located in the face of the cabinet within 12 inches below the counter top (the *NEC* limit for such placements), cords will be draped over the counter edge, providing children a ready means to pull over an appliance that might cause them injury. See page 74 for locations where GFCI protection is required.

Install a range receptacle (discussed on pages 168–169) in the beginning, because installing it later will be much more costly and difficult.

Bedrooms   Some people want a ceiling light in each bedroom; others do not. If you install a ceiling light, have it controlled by a wall switch near the door. If there will be no ceiling light, be sure some of the receptacles are controlled by a wall switch. The "6 foot" rule will probably require at least four receptacles. Plan location carefully; one should be permanently accessible for the vacuum cleaner, and another near the bed for items such as reading lamp, clock radio, and heating pad. If the overhead outlet will be used to support a paddle fan, for maximum flexibility use the special boxes covered under this topic in Chapter 9.

Closets   Don't overlook closet lights. Install the fixture on the ceiling or on the wall above the door so it is separated from the storage space by 12 inches, defined as at least 12 inches out from the walls, or the actual width of a shelf if greater than 12 inches, all the way to the ceiling. This is a safety measure. Many fires have been traced to clothing touching bare lamps in closets. The temperature of the glass bulb of an incandescent lamp is often higher than 400° F. In no case can a pendant fixture or an incandescent lamp without an enclosure be installed in a closet. If the fixture is a recessed incandescent with a solid lens, or a surface-mounted or recessed fluorescent fixture, the distance from the storage space may be reduced to 6 inches.

Bathrooms   A ceiling light controlled by a wall switch is essential. One additional light above the mirror at the basin is often installed, but that is not good lighting. Two lights—one on each side of the mirror—are needed for applying makeup or for shaving. Fluorescent brackets are ideal for the purpose. Pendant fixtures, ceiling paddle fans, and track lighting are prohibited in the area within 3 feet horizontally and 8 feet vertically of the tub or the shower area. A GFCI-protected receptacle (See page 74) is required within 36 inches of the outside edge of each basin. A dedicated 20-amp circuit is required for all the bathroom receptacles, unless the circuit (which still must serve only bathroom loads) supplies only one bathroom. In this case, the 20-amp circuit may serve the lighting outlets as well as the receptacle within the one bathroom. Thus, you may supply any or all the bathroom receptacles (in however many bathrooms) on a 20-amp circuit, or all the load, but in only a single bathroom, on a similar circuit.

Stairs   Where there are stairs, locate a fixture where it will light every step. Use three-way switches so the light can be turned on and off from both the top and bottom of the stairs. This is inexpensive insurance against accidents, and it is an *NEC* requirement if the stairs have six risers or more. If there is an intermediate landing with an entryway, as in many split-level houses, an additional switch must be placed at the intermediate landing.

Hallways, entrances, and basement   Every hallway deserves a light. For a long hallway, install three-way switches to control the light from either end. Install a receptacle for a vacuum cleaner. At each entrance to the house, outdoor lights controlled by switches indoors are essential. A motion-sensitive light is an option for your outdoor lighting. When people approach the door from outside at night, the light will turn on automatically. This is especially useful if the house does not have an attached garage. In the basement, install numerous receptacles for washer and dryer, workshop motors, and similar uses. If the basement has a finished room, the receptacles in that room follow the normal rules for habitable rooms, and at least one additional receptacle must be located in the unfinished portion. If the finished portion divides the basement into two or more unfinished areas, receptacles need to be provided for each unfinished area.

GFCI protection is required for receptacle outlets in unfinished basements. The former exceptions were deleted in the 2008 *NEC* cycle. Formerly exempted receptacles that aren't readily accessible now require GFCI protection. In addition, receptacles in dedicated space supplying appliances that aren't easily moved, including laundry equipment, now require GFCI as well. For example, a receptacle for a piped-in-place basement sump pump is now included in the GFCI protection requirement. Receptacles supplying alarm systems are still exempted, *but only for alarm receptacles that feed a fire alarm control panel.* In the case of an alarm system transformer, you will often find a mounting bracket designed to be secured to the cover mounting screw in the middle of a duplex receptacle. In this case, ask the inspector whether he or she would accept a conventional duplex receptacle with only the bottom receptacle connected (as shown in Fig. 10–12 but with both side tabs broken out).

CHOOSING LIGHTING FIXTURES

Good lighting requires not only sufficient light but also proper distribution of light throughout the room. It is easier to read in the shade of a tree than in direct sunlight even if there is less light in the shade. In direct sunlight, the light all comes from one point—the sun. In the shade of a tree, it comes from all directions. Similarly, fixtures with exposed lamps produce light from one point, causing sharp shadows and glare that make reading or sewing or other close work difficult. Provide well-diffused "shade of the tree" light by using fixtures that do not have exposed lamps.

For best lighting, ceilings should be a light color. A white ceiling reflects most of the light thrown against it. Ivory is not quite as good for the purpose. A flat finish is better than a glossy finish.

Some fixtures provide good illumination and others are primarily decorative. If you have good floor and table lamps to provide plenty of lighting where people sit, there is no reason why fixtures in the same room cannot be the decorative kind that provide some general lighting but not enough for exacting work. When ceiling fixtures are to be the primary source of light in a room, take care to locate them where they will provide lighting for critical seeing tasks. Buy the styles of fixtures that suit your taste and budget. Make sure the fixtures you select are electrically good and safe—look for the listing label on each fixture. Good lighting is not necessarily expensive. Lighting fixtures are available in a wide range of prices, and in many places a 100-watt lamp can be burned for a penny an hour.

Lamps (light bulbs)   Bulbs? Lamps? Different people call them by different names. Lamps are commonly bought and sold as "light bulbs" but technically only the outer glass globe is a bulb, while the entire unit is a lamp. In this book, they are called lamps in accordance with the terminology used in the *NEC*. See Fig. 6–1 for the symbol used to indicate a lamp in diagrams used in this book.

Types and efficiency   There are many types of lamps: incandescent, fluorescent, metal halide, mercury vapor, sodium vapor, etc. One 100-watt lamp gives 15 percent more light than three 40-watt lamps (120 watts). In other words, a fixture that uses one large lamp will in general provide more light than one using several small lamps of equivalent total wattage. However, incandescent lamps are inherently inefficient since they produce heat primarily; for this reason their availability in common sizes is being phased out by law. Replacements often use high-efficiency LED lamps.

Labeling   Federal regulations require most of the commonly used incandescent and fluorescent lamps to be labeled with the light output measured in lumens, energy used measured in watts, and the life in hours. To save energy, choose a lamp based on the light output and then select the lowest wattage lamp that will deliver that light.

Incandescent lighting   Ordinary lamps have an average life of 750 to 1,000 hours if used on circuits of the voltage stamped on the lamps, which is the voltage for which they were designed. There are also several newer types on the market.

"Long life" incandescent lamps   It is a simple matter to increase the life of a lamp by burning it at a voltage lower than it was designed for, as is done in "long life" lamps. For example, a lamp designed for 135 volts but burned at 120 volts will last about four times longer; a 140-volt lamp burned at 120 volts will last about eight times longer. But there is a catch—the light per watt is reduced by 20 to 30 percent and the cost is increased by 25 to 35 percent. In addition, such "long life" lamps sometimes consume more watts than the number stamped on them. If such a lamp is marked "60 watts" but actually consumes 70 watts, it will seem to produce just as much light as a standard 60-watt lamp with a 1,000-hour life. Such "extended service" lamps can be useful if located where replacement is difficult. Some lamps have reduced wattages. For example, replacing a 75-watt lamp with a 67-watt lamp will certainly save energy—but it will also give less light.

Energy-efficient incandescent lamps   There have been some real advances in incandescent lamp efficiency. Bulbs filled with krypton gas instead of nitrogen reduce energy consumption by 10 to 20 percent with no noticeable decrease in lamp life or light output. The lamp-within-a-lamp has a tungsten-halogen lamp inside a compact quartz tube within a conventional lamp bulb. The halogen gas redeposits tungsten particles back onto the filament, reducing blacking and providing up to three times the life of a conventional incandescent lamp.

Although the initial cost is higher, these energy-efficient lamps provide a return on investment of up to five times over their life. They last longer and use less electricity, saving you money and conserving energy at the same time. They come in 42-watt, 52-watt, and 72-watt sizes to replace standard 60-watt, 75-watt, and 100-watt incandescents. Their average life is 3,500 hours.

Fluorescent lighting   Fluorescent lighting has many advantages over incandescent. It produces far more light per watt of power used, and fluorescent lamps last much longer than ordinary lamps. The life of fluorescent lamps depends mostly on the number of burning hours per start. If turned on and left burning continuously, they will last at least eight to fifteen times as long as ordinary lamps. Even if turned on and off frequently, the fluorescent will last three to five times as long as an ordinary lamp. Fluorescent lamps used in homes come in lengths from 18 to 60 inches, consuming from 15 to 60 watts. The 48-inch/32-watt is the most common.

Brightness and whiteness   A further advantage of fluorescent lamps is that their surface brightness is much lower than that of filament lamps, providing light that is more diffused and thus more comfortable. Standard fluorescent lamps come in different kinds of "white," including deluxe warm white, warm white, white, cool white, and deluxe cool white. The cool white is most commonly used. It produces light most like natural light and emphasizes the blue and green colors in objects lighted. The warm white produces light more like that of ordinary filament lamps and emphasizes red and brown. The deluxe varieties emphasize these qualities still more but are a bit less efficient, producing a little less light per watt consumed.

Energy-efficient fluorescent lamps There are energy-efficient lamps and ballasts, including solid-state ballasts, that have ratings other than "standard," and these are truly more efficient than the lamps and ballasts they replace. Compact fluorescent lamps as small as 7 watts can replace incandescent lamps in many applications at great savings in energy. These compact lamps are made with a self-contained ballast and an Edison screw-shell base so they can very simply replace an incandescent lamp. A compact fluorescent rated around 25 watts replacing a 100-watt incandescent ranges in price from $10 to more than $20. A 15-watt compact fluorescent lamp replaces a 60-watt incandescent, a 20-watt fluorescent replaces a 75-watt incandescent, and a 25-watt fluorescent has light output equivalent to a 100-watt incandescent. These fluorescent lamps have an average life of 10,000 hours. Another type is the two-piece circular compact fluorescent lamp that has a reusable electronic ballast. It comes in 22-watt and 30-watt sizes to replace 100-watt and 150-watt incandescent lamps.

Compact fluorescent lamps come in many different styles and shapes. Newer types are labeled "electronic" because of the type of ballast used. Some have large bases that may not fit within the harps of your table and floor lamps, so measure the harp width before your trip to the hardware or home center store where the compact fluorescents are available. (Some electric power suppliers also sell them.) Generally a square style will fit in a smaller space than a circular shape. Some lamps cannot be used with dimmers, electronic timers, or photocell devices. Read the package before purchase to make sure you can use the lamp for your intended purpose.

Operation The fluorescent lamp produces light by activating a coating of phosphors on the inner wall of the bulb by means of ultraviolet energy generated by a mercury arc. This tubular hot cathode fluorescent lamp, also called a low-pressure electrical discharge source, consists of a glass tube capped at either end with a coiled wire filament called an electrode, or cathode. The inside surface of the tube is coated with a phosphor powder coating, while the tungsten wire electrodes have a coating (called an emission mix) that increases their ability to emit electrons. Figure 2–3 shows the construction.

Light generation in a fluorescent lamp involves a series of operations. When the lamp is first energized, a starting voltage source establishes an electric arc—a high

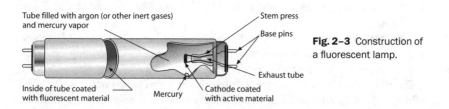

Tube filled with argon (or other inert gases) and mercury vapor

Stem press

Base pins

Exhaust tube

Inside of tube coated with fluorescent material

Mercury

Cathode coated with active material

Fig. 2-3 Construction of a fluorescent lamp.

velocity electron flow—through the tube. The arc interacts with the mercury vapor and other gases, generating ultraviolet light. This light strikes the phosphors on the inside of the tube and is then re-emitted in the form of visible light, the exact color being a function of the phosphor mix in use.

Once a lamp is started, the gas atmosphere offers little or no electrical resistance, so an auxiliary device called a *ballast* serves to limit the amount of current passing through the lamp's arc. The magnetic type ballast is essentially a choke coil with some other features. In the 1980s, improved ballasts were introduced for fluorescent lamps using solid-state electronic components in place of the traditional core and coil transformer operating at 60Hz, greatly boosting lighting system efficiencies. An electronic ballast optimizes the shape of the input voltage waveform sent to the lamp and uses frequencies in the 20 kHz to 45 kHz range, which excite the phosphors to a higher degree. These systems also have the ability to interface with sophisticated occupancy and other controls such as a daylight dimming response to sunlight entering a window.

The antiquated "preheat" method of fluorescent lighting required an external switch that would interrupt the flow of current to the ballast, resulting in a high-voltage "kick" as the magnetic field collapsed. This started the arc, and the current that followed was moderated by the magnetic choke supplied by the ballast windings. This style of fixture is seldom seen today because of a very poor power factor and substantial energy inefficiencies.

Lamps generally, including fluorescent ones, have a number associated with their designation, describing their diameter in eighths of an inch. Fluorescent lamps were usually produced as "T12" series lamps, the number indicating a 1 1/2-inch diameter. However, if the phosphors are closer to the arc, efficiency can be improved in terms of lumens of light delivered per watt of energy expended. This is why modern fluorescent fixtures increasingly use T8 lamps (1-inch diameter), or even smaller T5 lamps. Make certain, however, that you use the style tube for which the ballast has been rated.

There are two basic approaches to starting the initial arc using ballasts. A rapid-start fixture uses a small amount of current to heat the filament so it will produce the required initial flow of electrons. The instant-start fixture does not preheat the filament. Instead, a much higher voltage is applied directly across the tube when starting, sufficient to ignite the lamp without preheating the filaments. There is an important tradeoff to consider when making this choice. The low-power heating current consumes a small but noticeable amount of energy that the instant-start fixture does not waste. On the other hand, an instant-start voltage pulse is much tougher on the filament, exerting more wear on it than a rapid-start design. Therefore, if the fixture will be subject to repeated on-off cycles each day, choose a rapid-start design, or its modern electronic equivalent, the "programmed start" ballast. If, when started, it will generally be left on for the day, then choose an instant-start design.

The wiring diagrams in Figs. 2–4 and 2–5 illustrate the differences. The rapid-start model has current flowing through the filaments, and the instant-start model

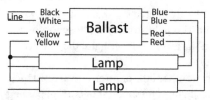
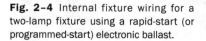
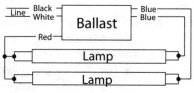
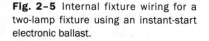

Fig. 2–4 Internal fixture wiring for a two-lamp fixture using a rapid-start (or programmed-start) electronic ballast.

Fig. 2–5 Internal fixture wiring for a two-lamp fixture using an instant-start electronic ballast.

has both filament leads brought together, since current will not be used to pass from one end of the filament to the other. These designs allow for a single, traditional, bi-pin-end fluorescent tube to be used in either fixture. Note that some instant-start fixtures, especially older T12 systems (including the older 8-foot instant-start models), use only a single-pin contact on each end of the tube. These tubes can only be used in instant-start fixtures. The two-lamp ballast shown in Fig. 2–5 also shows that the two tubes are wired in parallel, so if one lamp fails, the other lamp can remain lit. Three- and four-lamp ballasts are also available.

LIGHT EMITTING DIODE (LED) LUMINARIES

Luminaires that employ LED technology are making steady inroads into the market because of their inherent energy efficiency. They convert electrical energy directly into light through electrical effects on certain semiconductive rare-earth elements. This removes the intervening step of applying relatively large amounts of heat (incandescent) or an electrical arc (fluorescent). However, their electronic drivers do generate heat, so as always there is no such thing as a completely free lunch. Their components are also expensive, resulting in relatively high start-up costs.

These start-up costs are frequently capable of mitigation for large scale applications by utility and governmental grants designed to improve the overall energy budget. In addition, LED-driven replacements for conventional Edison-base incandescent lighting have greatly expanded the market penetration of this lighting method, having become commonly available on the shelves of home improvement stores and other large-scale merchandisers.

These replacement lamps can run in the range of twenty to thirty times that of an incandescent lamp. However, those costs are justifiable provided the components provide the advertised lifespan. This technology is here to stay, but there is still a ways to go to assure cost-effectiveness with respect to reliable lifespans in real-world applications.

LIGHTING AND RECEPTACLES FOR GARAGES AND OUTBUILDINGS

If the garage is attached to the house, treat it as another room as far as lighting is concerned. Three-way switches are preferred for convenience, one at the garage door and the other at the door between garage and house. If there is an entry door for personnel, add a wall-switch controlled outside light (This need not be done at the vehicle door). Detached garages follow the same rules if wired. Most people will want a light that can be controlled from both the garage and the house, requiring a three-way switch at each location.

NEC 210.52(G) requires every garage (if wired) to have at least one GFCI-protected receptacle outlet, and one for each space in multicar garages. More than one may also be desirable for a battery charger, tools, etc. In garages and outbuildings, almost all receptacle outlets are required to be GFCI-protected, and this includes outlets that are not readily accessible, such as a ceiling outlet for a garage door opener, and outlets for appliances in dedicated space, such as a freezer. Most outlets must be GFCI-protected. In a garage, it is convenient to have receptacle outlets that are always live rather than receptacles that are turned off with the garage light. With receptacles permanently on, you can plug in a battery charger and have it work all night without having the garage light on. See pages 120–123 for procedures and wiring diagrams for detached buildings.

Chapter 3
MEASURING ELECTRICITY

WATER IS MEASURED IN GALLONS, wheat in bushels, meat in pounds. Electricity cannot be poured into a measure or weighed on scales, but rather is something that must be considered as always in motion.

UNITS OF MEASUREMENT

We need to measure how much electricity flows past a given point at a given moment or in total over a period of time. To arrive at a measurement of electricity, the rate at which a quantity of it flows (coulombs per second, or amperes) and the pressure it is under when it flows (volts) are combined to arrive at wattage.

Amperes The absolute measure of a quantity of electricity is the "coulomb." We can speak of electricity in terms of "coulombs per second" in the way we might speak of water in motion in terms of gallons per second. Few people may recall ever hearing the word "coulomb," because instead of "coulombs per second" we use the simpler term "ampere" (abbreviated "amp"). One ampere of electricity is defined as current flowing at the rate of one coulomb per second. (Note that we don't say "amperes per second," but just "amperes.")

An electric current is the flow of electrons past a given point. One ampere is the movement of the charge associated with 6.28 billion billion (6,280,000,000,000,000,000) electrons per second. Wires are not electron hoses, and what actually transfers the energy we rely on is the electromagnetic field that is around and through the conductor, guiding it on its way to doing useful work.

Volts Water, air, and other substances can be put under pressure, which we commonly speak of in terms of pounds per square inch. Electric power is under pressure that is measured in volts (abbreviated "V"). Any ordinary dry cell or flashlight cell, when new, develops a pressure of about 1½ volts. One cell of a car battery develops 2 volts; six cells together develop 12 volts. Most house and farm wiring is at 120 volts for lighting and 240 volts for permanently installed appliances and for motors that run machinery. The voltage at which power is transmitted over high-voltage lines varies from 2,400 volts for short distances to 500,000 or more volts for long distances.

Watts and kilowatts   The total amount of power in a circuit at a given moment is expressed in "watts." Amperes alone or volts alone do not tell us the actual amount of power in a circuit. Both must be considered. Amperes and volts together tell us how much power is in a circuit at a given moment: volts × amperes = watts.

This formula is always correct with direct current, but it is correct only part of the time with alternating current. It is correct with lamps, ranges, toasters, and similar heating appliances. It is not correct with motors, or loads with transformers (radio, TV) or with ballasts (fluorescent lamps)—in all these the watts are somewhat less than volts × amperes. For ac load calculations, the *NEC* uses volt-amperes (VA) instead of watts.

Any wattage may consist of either a low voltage and high amperage, or a higher voltage and lower amperage. A lamp drawing 5 amps from a 12-volt battery consumes 60 watts (5 × 12 = 60); another lamp drawing ½ amp from a 120-volt line also consumes 60 watts (½ × 120 = 60 watts). The voltage and amperage differ widely, but the actual watts or power consumed by the two lamps is the same.

Watts measure power just as horsepower does. As a matter of fact, 746 watts is equal to 1 hp. A motor that delivers 1 hp delivers 746 watts and could just as well be called a 746-watt motor. (It uses more than 746 watts because some power is wasted as heat, and it also takes some power to run the motor even when it is not delivering power). A lamp that uses 746 watts could just as well be called a 1-hp lamp.

A watt is a very small amount of power. When speaking of large amounts of power, it is simpler to speak of kilowatts (the Greek word "kilo" means thousand). One kilowatt (abbreviated "kW") is 1,000 watts. We speak of watts, not watts per hour or kilowatts per hour, just as we say that an automobile engine delivers 200 hp, not 200 hp per hour.

Watthours and kilowatthours   Watts and kilowatts measure the rate at which power is being used at any given moment. Watthours and kilowatthours (units of energy) measure the total amount of power that has been used during any specified interval of time. One watt used for 1 hour is 1 watthour. Multiplying the watts used by the number of hours gives the watthours. A 60-watt lamp used for 6 hours consumes 360 watthours (60 × 6 = 360). A 2,000-watt room air conditioner used for 2 hours consumes 4,000 watthours (2,000 × 2 = 4,000).

A watthour is a very small amount of energy, so it is common to speak of kilowatthours. A kilowatthour is 1,000 watthours. The air conditioner mentioned in the previous paragraph, consuming 4,000 watthours in 2 hours, consumes 4 kilowatthours (abbreviated "kWh"). Electric energy from your power supplier is measured and paid for by the kilowatthour.

One kilowatthour will operate the average clothes washer for about 3 hours. It will operate a 1-hp motor for about an hour or pump about 1,000 gallons of water. It will operate the average stereo about 15 hours, a 50-watt lamp for about 20 hours, or an electric clock for about 3 weeks. One kilowatthour costs from 6 to 12 cents depending on your location.

RESIDENTIAL ELECTRICAL POWER

The amount of electricity that you and other consumers of electric power use leads to the cost you pay for electric service from your power company. You can check your consumption on your electric meter (installed by your power company) and on your monthly statements. Methods to reduce costs include using energy-saving lamps (discussed on pages 15–16) and energy-efficient appliances (discussed on page 163). Examples of power consumption by various household appliances are given in Table 3–1.

Reading your meter   Some meters provide a digital readout of the total kilowatthours used by the household. You read a digital meter just as you do the odometer on your car.

Older meters have dials as shown in Fig 3–1. Two of the pointers move in one direction, the others in the opposite direction. Four dials are shown, but you may have five dials on your meter. To read the dial meter: Assume it is the beginning of the month; going from left to right simply write down the number that each pointer has passed, as on a clock. The total of the meter in Fig. 3–1 is 1,642 kWh.

Fig. 3–2 shows the same meter a month later. One of the pointers points directly to the 7. Before writing down 7, look at the pointer on the dial to the right; it has not quite reached the zero. Therefore, even though the pointer seems to point directly to 7, it has not actually reached the 7, so write down 6 instead, for a total reading of 2,269 kWh. (If the last pointer were just past the zero, the total would be 2,270.) The difference between the two readings, 627 in this case, represents the number of kilowatthours of electricity used during the month.

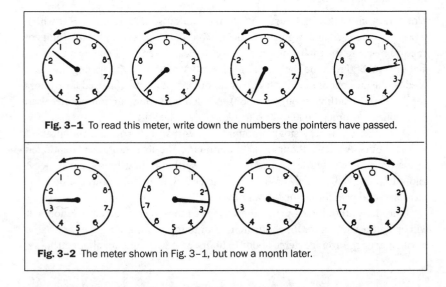

Fig. 3–1 To read this meter, write down the numbers the pointers have passed.

Fig. 3–2 The meter shown in Fig. 3–1, but now a month later.

Power rates   The rate charged for domestic electric power averages about 8.5¢ per kilowatthour in the United States, but the rate varies greatly. Before the energy crisis of the 1970s, rate schedules were stepped so the more power you used, the lower the average cost per kWh. Now, to encourage conservation of energy, most rates are flat after an initial step, and some are inverted—the more power you use, the higher the average cost per kWh.

Here is an example of an inverted rate structure:

First 100 kWh used per month....................4.8¢ per kWh
All over 100 kWh used per month...............9.2¢ per kWh

The meter reading in Figs. 3–1 and 3–2 showed 627 kWh consumed during the month. The bill is figured this way:

100 kWh at 4.8¢ $4.80
527 kWh at 9.2¢ $48.49
627 kWh..Total $53.29
Average per kWh...8.5¢

Considering the rate of inflation and fluctuating fuel costs, these rates may not be typical at the time you are reading this.

Operating cost per hour   To find out how much it costs to operate any electrical load for one hour, multiply the watts that the item consumes by the rate in cents per kWh, placing the decimal point five places from the right to arrive at the cost in dollars per hour.

Some examples of operating costs:

ELECTRICAL LOAD	WATTS × RATE	OPERATING COST IN DOLLARS PER HOUR
60-watt lamp at 8.5¢ per kWh	60 × 8.5 = 510	$0.0051 (slightly more than a half-cent per hour)
600-watt appliance at 8.5¢ per kWh	600 × 8.5 = 5100	$0.051 (5.1¢ per hour)

How long does it take a particular electrical load to consume one kWh? To find out, simply divide 1,000 by the wattage of the load to get the number of hours, as in the following examples.

ELECTRICAL LOAD	1000 ÷ WATTS	TIME IT TAKES TO CONSUME ONE KILOWATTHOUR
40-watt lamp	1000 ÷ 40 = 25	25 hours
600-watt appliance	1000 ÷ 600 = 1.666	1.666 hours (about 1 hr 40 min)
electric clock using about 2 watts	1000 ÷ 2 = 500	500 hours

Watts consumed   The following table will aid you in estimating power required (watts consumed), or the operating cost, for various appliances. The figures are only approximate, and new energy-saving appliances are appearing all the time. Appliances often have their total wattage (representing the load on the circuit) listed on their nameplates.

Table 3-1   WATTS CONSUMED BY VARIOUS APPLIANCES

FOOD PREPARATION	Watts
Blender	500 to 1,000
Coffeemaker	500 to 1,200
Convection oven	1,500 to 1,700
Dishwasher	1,000 to 1,500
Food processor	500 to 750
Frying pan	1,000 to 1,200
Hot plate, per burner	600 to 1,000
Knife	100
Microwave oven	1,000 to 1,500
Mixer	120 to 250
Oven, separate	4,000 to 5,000
Range	8,000 to 14,000
Range top, separate	4,000 to 8,000
Roaster, large	1,440
Rotisserie (broiler)	1,200 to 1,650
Toaster	500 to 1,200
Toaster oven	1,550
Trash compactor	1,250
Waste disposer	500 to 900

FOOD STORAGE	Watts
Freezer, household	300 to 500
Refrigerator, non-frostless	150 to 300
Refrigerator, frostless	400 to 600

MOTORS	Watts
¼-hp	300 to 400
½-hp	450 to 600
Over ½-hp, per hp	950 to 1,000

PERSONAL CARE	Watts
Hair dryer	350 to 1,600
Heating pad	50 to 75
Shaver	8 to 12
Toothbrush	20 to 50

OFFICE EQUIPMENT	Watts
Computer, monitor, printer	200 to 600
Fax machine	200

ENVIRONMENTAL COMFORT	Watts
Air conditioner, central	2,500 to 6,000
Air conditioner, room	800 to 2,500
Blanket	150 to 200
Dehumidifier	250
Fan, attic	800
Fan, bathroom ventilator	500
Fan, ceiling	40 to 300
Fan, whole house	1,000
Fan, portable	50 to 200
Heat lamp (infrared)	250 to 500
Heater, portable	1,000 to 1,500
Heater, wall mounted	1,000 to 4,500
Heater, water bed	800
Heating, electric forced air	10,000
Humidifier	450

ENTERTAINMENT	Watts
Compact disc player	10 to 25
Projector, slide or movie	300 to 500
Stereo	30 to 100
TV (up to 46 in)	150 to 260
VCR	40 to 70

LAUNDRY	Watts
Dryer	4,000 to 6,000
Iron	600 to 1,200
Washer	300 to 800
Water heater	2,000 to 5,000

MISCELLANEOUS	Watts
Clock	2 to 3
Hot tub (electrically heated)	12 kW
Lamps, fluorescent	7 to 60
Lamps, incandescent	10 upward
Sewing machine	60 to 90
Spa (electrically heated)	12 kW
Swimming pool heater	12 kW
Vacuum cleaner	250 to 1,200

TYPES OF ELECTRIC CURRENT

Current is either alternating or direct. On batteries, one terminal is always positive (+) and the other terminal is negative (–). The type of current characterized by each wire being always of the same polarity—either positive or negative—is known as direct current (dc). Current from a battery is always dc. The current coming into your home or farm is alternating current (ac).

Alternating current (ac)   In alternating current each wire changes or alternates continually between positive and negative. The change from positive to negative and back again to positive is a "cycle." This takes place 60 times every second, and such current is known as 60-hertz or 60-Hz current. The term "hertz" is used instead of "cycles per second." It is named for the German scientist Heinrich Hertz who discovered the cyclical nature of electrical waves.

Sixty times every second, each wire is positive; and 60 times every second, it is negative; and 120 times every second, there is no voltage at all on the wire. The voltage is never constant but is always gradually changing from zero to a maximum of usually about 170 volts, but averaging 120 volts; and such current is known as 120-volt current (See Fig. 3–3).

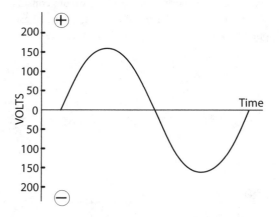

Fig. 3–3 One cycle of 120-volt alternating current. If it is 60-Hz current, all the changes shown take place in ⅟₆₀ second.

You might wonder why lights do not flicker if there is no current in the wire 120 times every second. The filament in the lamp does not cool off fast enough to create flickering, but very small lamps used on 25-Hz current (where there is no voltage on the wire 50 times every second) do have an annoying flicker.

Single-phase and three-phase current   The current described in the previous paragraph is single-phase current. Remember that if 120-volt, 60-Hz current flows in a pair of wires, 120 times every second the wires are dead (no voltage at all); 120 times every second the voltage is about 170 volts; at all other times it is somewhere in between, but averaging 120 volts. The voltage is always changing, but the changes are so rapid that for most purposes it can be considered a steady 120-volt current. Single-phase power may be supplied with either two or three wires from the power supplier's transformers (See Fig. 7–1 and related text).

Three-phase power is used in factories and commercial establishments where there are many motors. It is seldom found in homes or farms. To understand how three-phase power is generated, imagine three separate electric generators all on a single shaft, arranged so the voltage reaches its maximum at different times in each of the three generators: first in one, then in the second, then in the third, then again in the first, and so on. A pair of wires would run from each of the three generators. The three generators together are said to deliver three-phase power (although the power from any one generator is still single-phase). In actual practice, the three generators become a single generator with three coils of wire (windings); the three pairs of wires become three wires.

Three-phase power requires three-wire instead of two-wire high-voltage transmission lines and three transformers instead of one. (Three-phase power may be supplied with either three or four wires from the transformers.) Do not be misled into thinking that because there are three wires in the service entrance, the result must be three-phase power. On farms and in homes, the presence of three wires almost always means single-phase, three-wire, 120/240-volt service. If you are fortunate enough to have three-phase power available, by all means use three-phase motors for greater efficiency. If you do have a 240-volt, three-phase supply, consult the power supplier to determine whether you also have 120/240-volt, single-phase power for lighting and small appliances without installing your own transformer.

PART 2 WIRES, CIRCUITS, AND GROUNDING

Chapter 4
WIRE—SELECTING AND CONNECTING

WIRES ARE USED AS CONDUCTORS. A conductor is any material that can carry the flow of electric current. The terms *wire* and *conductor* are used interchangeably in this book. *Insulators* are materials that do not conduct electric current. Metal wire is enclosed in insulating material such as plastic to help protect against stray current.

Electricity flows more easily in some materials than others. Copper wire is the best material for ordinary purposes. If iron wire were used, it would have to be about ten times as large in cross-sectional area as copper wire. Other conductors include cable and busbars. Humans and animals can accidentally become conductors resulting in electric shock, so always use caution when working with electricity.

All references in this book are to copper wire, except in the discussion of aluminum wire later in this chapter and in Chapter 19.

WIRE SIZES

Copper wire sizes are indicated by number–the larger the number, the smaller the wire. See Fig. 4–1. The most common size for house wiring is 14 AWG, which is not quite as thick as the lead in a pencil. Numbers 12, 10, and 8 AWG and so on are larger than 14 AWG; 16, 18, and 20 AWG and so on are progressively smaller. Number 14 AWG is the smallest size permitted for ordinary house wiring, and 1 AWG is generally the heaviest used in residential and farm wiring. Still heavier sizes are 1/0, 2/0, 3/0 and 4/0 AWG, the 4/0 being almost half an inch in diameter. Numbers 16 and 18 AWG conductors are used mostly in flexible cords and the still finer sizes are used mostly in the manufacture of electrical equipment such as motors. Number 18 AWG is also commonly used in wiring doorbells, chimes, thermostats,

| 2/0 | 1/0 | 2 | 4 | 6 | 8 | 10 | 12 | 14 | 16 | 18 |

Fig. 4–1 Approximate diameters of different sizes of copper wire, without the insulation.

and similar items operating at less than 30 volts. Wire of the correct size must be used for safety and efficiency as discussed below in terms of *ampacity* and *voltage drop*.

Ampacity   Ampacity is the safe carrying capacity of a wire as measured in amperes. When current flows through wire, it creates a certain amount of wasted heat. The greater the amperes flowing, the greater the heat. Doubling the amperes without changing the wire size increases the amount of heat four times. To avoid wasted power, a wire size that limits the waste to a reasonable figure should be used. Of even more concern, if the amperage is allowed to become too great, the wire may become hot enough to damage the insulation or even cause a fire. The *National Electrical Code* (*NEC*) is not concerned with wasted power, but it is concerned with safety; therefore it sets the ampacity—the maximum amperage that various sizes and types of wires are allowed to carry.

Conservative ampacities for common wire sizes are shown in Table 4–1. For larger sizes, consult *NEC* Table 310.15(B)(16). Insulation types are described under the "Wire Types" heading in this chapter. Equipment and devices (circuit breakers, switches, panelboards, etc.) have been tested, unless otherwise marked, at the ampacities in Column A of Table 4–1 up to 100 amperes and in Column B over 100 amperes, so even though the *NEC* tables might indicate the conductor has a higher ampacity, the termination may limit the ampacity to those shown.

Table 4–1 AMPACITY OF WIRES (Based on *NEC* Table 310.15(B)(16)]
Not more than three current-carrying conductors in conduits or other raceways, or in cable assemblies, or directly buried in the earth.

WIRE SIZE	Wire insulation types and temperature ratings					
	Copper			Aluminum (or copper-clad aluminum)		
	A	B	C	A	B	C
	60°C TW	75°C RHW, THW, THWN	90°C THHN, XHHW-2 THWN-2	60°C TW	75°C RHW, THW, THWN	90°C THHN, XHHW-2 THWN-2
14*	15	15	15	—	—	—
12*	20	20	20	15	15	15
10*	30	30	30	25	25	25
8	40	50	55	35	40	45
6	55	65	75	40	50	65
4	70	85	95	55	65	75
3	85	100	115	65	75	85
2	95	115	130	75	90	100
1	110	130	145	85	110	115
1/0	125	150	170	110	120	135
2/0	145	175	195	115	135	150
3/0	165	200	225	130	155	175
4/0	195	230	260	150	180	205

* In the *NEC*, the ampacities of these wires is higher than this, but a footnote reference to 240.4 (D) limits their overcurrent protection to the values listed here.

Column C covers conductors with high-temperature insulation (90°C). Although these wires are very common today, the termination rules tend to limit the usable current in these wires to the amounts in Columns A and B as noted.

Conductors have different ampacities for each variation in ambient temperature, proximity to other conductors, type of insulation, depth of burial, etc. Variations in ambient temperature are covered in the *NEC* by the correction factors in *NEC* Table 310.15(B)(2)(a). Adjustments to ampacity due to numbers of conductors in the same raceway or cable are covered in *NEC* 310.15(B)(3)(a). Ampacity calculations are some of the most complex in the *Code*; however, for simple home and light commercial work, the values in Table 4–1 should suffice.

Voltage drop   If forcing too many amperes through a wire only caused a certain amount of wasted power, we might consider it a mere nuisance and minor loss. However, it also causes voltage drop. Actual voltage is lost in the wire so that the voltage across two wires is lower at the end than at the starting point. For example, if you connect two voltmeters into a circuit, as in Fig. 4–2, one at the main switch and one across a 1-hp motor at a distance, you will find that the voltage at the motor is lower than at the main switch. The meter across the main switch may read 120 volts. If 14 AWG wire is used to the motor, the voltage across the motor terminals will be about 119 volts if the motor is 10 feet away, but only about 112 volts if it is 100 feet away. The difference is lost in the wire and is known as voltage drop. Voltage drop is wasted power, but there is another important consideration: appliances work very inefficiently on voltages lower than the voltage for which they were designed. At 90 percent of rated voltage, a motor produces only 81 percent of normal power and a lamp produces only 70 percent of its normal light.

In the example shown in Fig. 4–2, if 200 feet of 14 AWG wire is used, the drop is from 120 to 112 volts, or 8 volts, about 7 percent. If 12 AWG wire had been used, the drop would have been reduced about 60 percent to 3.2 volts, only about 2.5 percent of the starting voltage. The larger the wire, the less the voltage drop.

Voltage drop can't be reduced to zero, but it can be kept at a practical level by using wire of sufficient size. A drop of 2 percent is considered acceptable. If the starting point is 120 volts, 2 percent is 2.4 volts, so the actual voltage at the point where the power is consumed is 117.6 volts. If the starting point is 240 volts, the voltage at the point of consumption is 235.2 volts. The apparent savings in initial cost by using undersize wire is soon offset by the cost of power wasted in the wires and by the reduction in efficiency of lamps, motors, and so on.

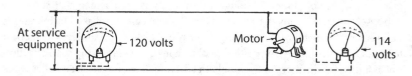

Fig. 4–2 Circuit showing voltage drop—voltage at motor is lower than at starting point.

Selecting wire size   Choose a type and size of wire that has an ampacity rating at least equal to the expected load (in amperes) and that is large enough to limit voltage drop to a practical range. The *NEC* permits nothing smaller than 14 AWG for ordinary wiring. It is better to consider 12 AWG the smallest, and this is required in a few places by local ordinance. Larger size wires may be required to minimize voltage drop in long runs, or to allow larger motors to start (wire size for motors is further discussed on page 185). If you need wire heavier than the minimum permitted, it is somewhat complicated to figure the right size but simple to look it up in tables. First determine the amperage to be carried by the wire. Most appliances will have a nameplate giving the total amperes or watts rating.

Tables 4–2 and 4–3 on the next page show one-way distances in feet for 2 percent voltage drop at 120 and 240 volts. These are recommendations in *NEC* 210.19(A) Informational Note No. 4 and 215.2(A)(1) Informational Note No. 2. Use the table that corresponds to the voltage of the circuit in question. To operate a load 300 feet away requires 600 feet of wire, but look for the figure 300 in the table. The distances under each wire size are the distances that size wire will carry the different amperages (or wattages) in the left-hand columns, with the customary 2 percent voltage drop. In the 120-volt table, to determine how far 8 AWG wire will carry 20 amps, follow the 20-amp line until you come to the 8 AWG column: the answer is 90 feet. If a distance is marked with an asterisk (*), it indicates that Type TW wire in conduit or cable, or a cable buried directly in the ground, may not be used because the amperage in the left-hand column is greater than the ampacity of Type TW. Select the proper type of wire from Table 4–1 on page 28, or from Table 310.16 in the *NEC*.

Compare the 120-volt and 240-volt tables. Note that at 240 volts, any given size of wire will carry the same amperage twice as far as at 120 volts with the same percentage of voltage drop. It will carry the same number of watts four times as far.

When wires are run outdoors overhead, they must be large enough to carry the amperage involved without excessive voltage drop. They must also be large enough to support their own weight. *NEC* 225.6(A) requires a minimum of 10 AWG for spans up to 50 feet and 8 AWG for larger distances. For distances over 150 feet, it is wise to use an extra pole. In northern areas where the wires often must support a heavy ice load, consider using a size larger than electrically required. If the wire is installed on a hot summer day, leave considerable slack; otherwise the contraction of the wires in winter may pull the insulators off your buildings.

Weatherproof wire has a covering of neoprene or impregnated cotton over the conductor. The *NEC* calls it "covered"—the covering is not recognized as insulation. Weatherproof wire must not be used for ordinary wiring, but only for overhead wiring outdoors. Although its ampacity is higher than that shown in Table 4–1 because it runs in free air, any equipment connected at its terminations falls under the general rules, resulting in the use of Columns A and B as a practical matter. Using the higher ampacities involves a level of engineering sophistication not assumed for the users of this book. In addition, don't forget that smaller wires lead to higher voltage drop.

Table 4-2 ONE-WAY DISTANCES FOR 2% VOLTAGE DROP BY WIRE SIZE AT 120 VOLTS SINGLE-PHASE

Amperes	Watts at 120 volts	14 AWG	12 AWG	10 AWG	8 AWG	6 AWG	4 AWG	2 AWG	1/0 AWG	2/0 AWG	3/0 AWG
5	600	90	140	225	360	570	910				
10	1,200	45	70	115	180	285	455	725			
15	1,800	30	45	70	120	190	300	480	765	960	
20	2,400	20*	35	55	90	145	225	360	575	725	915
25	3,000	18*	28*	45	70	115	180	290	460	580	730
30	3,600	15*	24*	35	60	95	150	240	385	485	610
40	4,800			28*	45	70	115	180	290	360	455
50	6,000			23*	36*	55	90	145	230	290	365

In the tables above and below, the figures represent *one-way* distances in feet, not the total wire length for two-way distances.

* In both tables, for distances marked with an asterisk (*) Type TW wires in conduit or cable may not be used because they do not have enough ampacity. For all distances marked with the asterisk, select a type of wire with sufficient ampacity (depending on whether in conduit or cable, or in free air) from Table 4-1 on page 28.

If you wish to permit 4 percent drop, double the distances shown. Multiply the distances by 2.5 to permit 5 percent drop.

Table 4-3 ONE-WAY DISTANCES FOR 2% VOLTAGE DROP BY WIRE SIZE AT 240 VOLTS SINGLE-PHASE

Amperes	Watts at 240 volts	14 AWG	12 AWG	10 AWG	8 AWG	6 AWG	4 AWG	2 AWG	1/0 AWG	2/0 AWG	3/0 AWG
5	1,200	180	285	455	720	1,145					
10	2,400	90	140	225	360	570	910	1,445			
15	3,600	60	95	150	240	380	610	970	1,530		
20	4,800	45*	70	115	180	285	455	725	1,150	1,450	
25	6,000	35*	55*	90	140	230	365	580	920	1,160	1,460
30	7,200	30*	48*	75	120	190	300	480	770	970	1,220
40	9,600		36*	56*	90	140	230	360	575	725	915
50	12,000			45*	70*	115	185	285	460	580	725
60	14,400				60*	95*	150	240	385	485	610
70	16,800				50*	80*	130	205	330	410	520
80	19,200					70*	115*	180	285	360	460
90	21,000					60*	100*	160	250	320	405
100	24,000					55*	90*	145*	230	290	365
125	30,000						75*	120*	190	240	300
150	36,000							95*	150*	195*	245
200	48,000							70*	115*	145*	185*

WIRE TYPES

Wire of the correct type as well as of the correct size must be used to assure a safe installation. Wires covered with various types of insulation are used for wiring the interiors of buildings. (Wires for outdoor installations are discussed at the bottom of page 30 and on pages 194–195 in the chapter on farm wiring.) Wire names indicate the type of insulation, and these names are generally abbreviated.

The different colors of wires indicate function. The special uses of white and green wires are summarized on page 53 and page 67. Black wires are "hot" (energized)—they carry current to electrical equipment. Red and blue, if used, are also hot; use of such additional colors makes it easier to tell wires apart in a complicated installation. Number 4 AWG and larger wires are usually available only in black, and weatherproof wire in all sizes is always black.

Number 10 AWG and smaller wire is usually solid—the copper conductor is a single solid strand. Number 8 AWG wire is usually stranded—several smaller wires are grouped together to make one larger wire (See Fig. 4–1), which is more flexible. This is true even for cabled makeups, although the *NEC* does allow 8 AWG to be solid if it is in the form of cable or is not to be drawn into conduit after installation. But wire that is 6 AWG or larger (with the exception of weatherproof wire) must always be stranded.

To protect them from damage, wires are used in two basic ways. They can be field installed in raceways, either by being pulled into tubular types called either conduits or tubing, depending on the specific wiring method, or by being laid into channels with removable covers. The other option is for them to be prefabricated by a manufacturing process, usually with some form of outer sheath or armor for physical protection. In addition to the wiring methods described in this book, many other types are available, but they are usually not used in residential and farm wiring. You will find them listed in your copy of the *NEC*.

Types TW, THW, and THHN (and/or THWN) wire   An older style wire was thermoplastic insulated without any outer layer. The conductor was covered by a single layer of plastic compound in a thickness relating to the size of the wire, and which strips off easily and cleanly. See Fig. 4–3. Type TW is moisture-resistant and suitable for use in wet locations. Type THW is resistant to both heat and moisture. Neither TW nor THW may be buried directly in the ground. The most common

Fig. 4–3 Type TW and its higher temperature relation THW (above) is seldom used, having been largely replaced by dual-rated THHN/THWN (below), which uses the same insulating compound, but in a much thinner layer overlaid by a tough outer layer of clear nylon that provides excellent mechanical protection. It may be used in wet or dry locations. Number 6 AWG or larger must be stranded, and 8 AWG must be stranded if pulled into conduit.

conductors have Type TW insulation, but thinner than on conventional Type TW wire, supplemented (for mechanical resistance) by a layer of nylon. If the wire is suitable for dry locations only, it is Type THHN, and for wet locations, Type THWN. Most wires of this type carry the dual designation THHN/THWN.

Type R wire　Formerly rubber-covered, this kind of wire has a synthetic polymer insulation, and it may have a moisture-resistant, flame-retardant outer covering of neoprene or PVC. Figure 4–4 shows the makeup of the original Type R conductor. Once the most popular of all kinds of wire, Type R is no longer used, and has been removed from the *NEC*. However the more modern higher-temperature versions, RHH (90°C) for dry and damp locations, and RHW (75°C) and RHW-2 (90°C) for dry or wet locations, are still used.

Fig. 4–4　Rubber-covered wire has rubber instead of plastic insulation, and it may have a fabric or other nonmetallic flame-retardant outer covering.

Aluminum wire　This is available in two types—aluminum and copper-clad aluminum (aluminum with a thin sheath of copper on the outside). Because aluminum is not quite as good a conductor, a larger size must be used than when using copper. A rule of thumb is to use aluminum two numbers heavier than copper: 12 AWG aluminum instead of 14 AWG copper; 4 AWG aluminum instead of 6 AWG copper, etc.

　　When aluminum wire was first used, it was connected to ordinary terminals that were suitable for copper, but it soon became evident they were not suitable for aluminum. The connections heated badly and led to loose connections, excessive heating, and sometimes fires. The product standards were revised, and test labs then required redesigned terminals, marked AL-CU, that were considered suitable for either copper or aluminum, but in the 15-amp and 20-amp ratings they were still not suitable for aluminum. The terminals were further redesigned and since 1971 only devices with the marking CO/ALR are acceptable. Note that terminals rated *higher than* 20 amps were not changed, so those marked AL-CU are still acceptable and they *must* be used when aluminum wire is installed.

　　The AL-CU or CO/ALR marks are stamped into the mounting yokes of switches, receptacles, and similar devices in order to remain visible without removal from the boxes in which they are installed. On larger equipment the marks are located so they remain visible after installation.

　　If your existing installation was made using aluminum, you would be wise to inspect all your receptacles and switches. If they are not marked CO/ALR, replace them all with devices that do have the mark, or reconnect them with copper pigtails. (Replacement instructions are on pages 221–222.)

　　Keep in mind that if the aluminum wire is copper-clad, any kind of terminal may be used. Push-in terminals, shown in Fig. 4–16, may be used with copper or copper-clad aluminum, but *not* with ordinary aluminum.

CABLE TYPES

Wires are often assembled into cables such as nonmetallic-sheathed cable or armored cable. When a cable contains two 14 AWG wires, it is known as 14-2 (fourteen-two) cable; if it has three 12 AWG wires, it is called 12-3, and so on. If a cable has, for example, two 14 AWG insulated wires and a bare uninsulated grounding wire, it is known as "14-2 with ground." If a cable has two insulated wires, one is always white and one black; if it contains three, the third is red.

Nonmetallic-sheathed cable   This is a very common type of cable containing two or three Type THHN or THHW wires. Many people call it "Romex," which is the trade name of one particular manufacturer. It is easy to install, neat and clean in appearance, and less expensive than other kinds of cable. Brief descriptions of the two kinds follow here. See Chapter 11 for a complete discussion of where and how to use nonmetallic-sheathed cable.

One kind of nonmetallic-sheathed cable is called Type NM by the *NEC*. As shown in Fig. 4–5A, the individually insulated wires are enclosed in an overall plastic jacket. (In older construction, occasionally still seen, the outer jacket was a braided fabric. In any case, the outer jacket must be moisture-resistant and flame-retardant.) Some manufacturers put paper wraps on the individual wires or over the assembly, although most modern assemblies only have paper over a bare grounding wire. Empty spaces between wires are sometimes filled with jute or similar cord. This type may be used only in normally dry locations, but never in barns on farms.

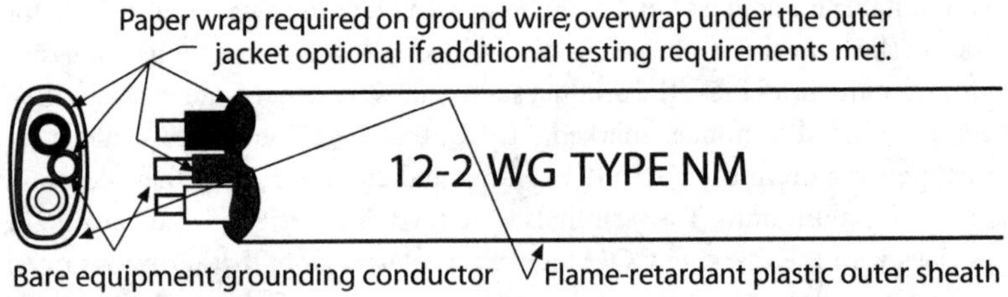

Paper wrap required on ground wire; overwrap under the outer jacket optional if additional testing requirements met.

12-2 WG TYPE NM

Bare equipment grounding conductor   Flame-retardant plastic outer sheath

Fig. 4–5A Nonmetallic-sheathed cable consists of two or more individual wires assembled into a cable. The Type NM two-wire with ground shown here may be used only in dry locations.

Solid plastic

12-2 WG TYPE NMC

Bare equipment grounding conductor

Fig. 4–5B Nonmetallic-sheathed cable, Type NMC, may be used in dry or damp locations.

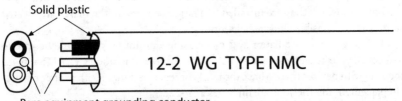

The other kind is called Type NMC by the *NEC* and is especially designed for damp or corrosive locations such as barns. It may also be used in ordinary dry locations. The individual insulated wires in Type NMC cable are embedded in a solid sheath of plastic material (See Fig. 4–5B). Sometimes there is a glass overwrap on each insulated wire. There is no fibrous material such as paper wraps or jute filler that can be affected by moisture as in ordinary Type NM.

Metal-clad cable   *NEC* Type MC (Fig. 4–6) has multiple conductors made up with an interlocking spiral armor of either aluminum or steel for physical protection. It is also available with a ribbed, seamless, flexible aluminum armor (the "corrugated" style) and a comparatively inflexible smooth aluminum armor, but the interlocking armor is the style usually seen. The corrugated and smooth constructions can be used in wet locations. It is also available with a nonmetallic jacket over the armor, and in that form the interlocking type can be used in wet locations, including direct burial if so marked on the cable. The individual wires may be type TW, THW, or THHN/THWN, the most common type used. This cable is also available with a bare aluminum grounding conductor running just under the spiral armor.

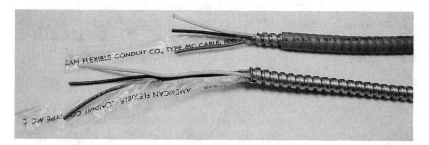

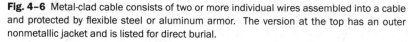

Fig. 4–6 Metal-clad cable consists of two or more individual wires assembled into a cable and protected by flexible steel or aluminum armor. The version at the top has an outer nonmetallic jacket and is listed for direct burial.

Type MC cable has largely supplanted its predecessor, armored cable (Type AC). This cable also has a spiral, interlocking armor, but it has a bonding wire that is tightly compressed against the inner edges of the convolutions by a paper wrap over the circuit conductors. This allows for the armor to have sufficient conductivity such that it qualifies as a grounding conductor without an additional green wire. It is for use only in permanently dry locations. Some major manufacturers have muddied this distinction, however, by producing an interlocking-armor metal-clad cable that has a full-sized aluminum grounding wire in the makeup. This wire is arranged outside the spiraled circuit conductors and in tight contact with the inner margin of the interlocking armor. Since the combination of this bonding wire and the armor has been evaluated as a grounding path, the bare wire usually does not have to enter an enclosure, just as the bonding strip of the armored cable does not have to enter enclosures. Details for using these cables are provided in Chapter 11.

Underground cable   Two special types of cable, USE and UF, are used for underground wiring. Type UF cable is very similar to the Type NMC mentioned previously and is commonly substituted for it. They are described in the chapter on farm wiring on page 195.

FLEXIBLE CORDS

Flexible cords are used to connect lamps, appliances, and other loads to outlets. Each wire consists of many strands of fine wire for flexibility. Over the wire is a wrapping of cotton to prevent the insulation from sticking to the copper. There are many kinds of flexible cords with varying kinds and thicknesses of insulation, depending on the purpose of the cord. The more common kinds are described here.

Type SPT-2 is the most common cord used for radios, floor lamps, and similar loads. As shown in Fig. 4–7, it consists of copper wires embedded directly in plastic insulation. It is tough, durable, and available in various colors. The same kind of cord with a rubber-like insulation is known as Type SP-2. These cords are commonly available only in 18 AWG and 16 AWG.

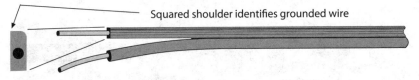

Squared shoulder identifies grounded wire

Fig. 4–7 In Type SPT-2, the wires are embedded in plastic. The cord is durable and attractive.

Type S, in Fig. 4–8, is a really durable cord that stands up to hard use. Each wire has rubber or plastic insulation. The two wires are bundled into a round assembly with jute or paper twine filling the empty spaces. Over all is a layer of tough, high-grade, rubber-like plastic. Type SJ is similar but with a thinner outer layer. Type SV or SVT are very similar but more flexible cords used on vacuum cleaners.

If the outer jacket is made of oil-resistant thermoset compound, the cord becomes oil-resistant and the designations become SO and SJO instead of S and SJ. If the outer jacket is a thermoplastic, then a "T" is added, as in ST or SJT, and if it is oil resistant, an "O" is added as well, as in STO or SJTO. If the inner wires are also oil resistant, then a second "O" is used, as in STOO.

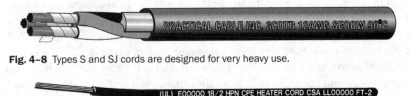

PRACTICAL CABLE INC. 3CDTR 12AWG SEOOW 90°C

Fig. 4–8 Types S and SJ cords are designed for very heavy use.

(UL) E00000 18/2 HPN CPE HEATER CORD CSA LL00000 FT-2

Fig. 4–9 Type HPN cord is used on irons, toasters, and other heating appliances.

"Heater cord" is used for irons, toasters, and other heating appliances. Heater cords formerly had asbestos under a cotton braid as part of the insulation system. Asbestos is no longer used as wire insulation. Figure 4–9 shows Type HPN heater cord; the wires are embedded in neoprene.

TOOLS USED FOR WIRING JOBS

Listed below are tools commonly used in wiring procedures. Tools are mentioned in the text where procedures are described. Also look in the index under "Tools" or individual tool names.

- Cable cutter for cutting armored cable
- Cable ripper for removing the outer jacket from nonmetallic-sheathed cable
- Conduit bender for bending conduit
- Hacksaw for cutting armored cable, EMT, and openings in existing walls
- Putty knife sharpened to lift floorboards to get at ceiling spaces from above
- Keyhole saw for cutting floorboards
- Fish tape for pulling wires into place
- Parallel-jaw pliers for use with connectors
- Pliers with crimping die for use with connectors
- Diagonal-cutting pliers ("dikes" in the trade) for cutting wire
- Wire stripper or knife for removing insulation from wire
- Test light for receptacles and fuses
- Continuity tester for testing that hot wires in switches are not accidentally grounded and for detecting broken wires in cords

TERMINALS FOR CONNECTING WIRES TO DEVICES

Various kinds of terminals provide the means for connecting wires to devices. Before wires can be connected to a device or spliced to another piece of wire, the insulation must be removed. The procedure begins with preparing the wire.

Removing insulation from wire   Using a tool called a wire stripper is the most convenient way to remove insulation from wire.

If you don't have a wire stripper, you can use a knife. Do not cut the insulation off sharply, as shown at A of Fig. 4–10, because it is too easy to accidentally nick the conductor, leading to later breaks. Hold your knife to produce an angle as shown in B of Fig. 4–10.

Always make sure the stripped end of wire is absolutely clean. Modern insulation

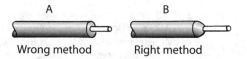

A   B

Wrong method   Right method

Fig. 4-10 Wrong and right methods of removing insulation from wires.

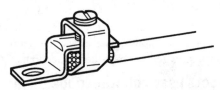

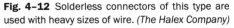

Fig. 4–11 Terminal on a typical receptacle or switch.

Fig. 4–12 Solderless connectors of this type are used with heavy sizes of wire. *(The Halex Company)*

strips off cleanly although some patience may be involved with larger stranded wires because the insulation often lays into the stranding intervals and must be teased out.

Two types of terminals There are two kinds of terminals. One kind consists of a terminal screw in a metal part with upturned lugs to keep the wire from slipping out from under the screw, as shown in Fig. 4–11. It may be used with 10 AWG and smaller wires, but it is very difficult to make a good connection if the wire is 10 AWG and stranded, or even smaller wires if finely stranded. The other kind is used mostly for wires larger than 10 AWG; the wires are inserted into the terminal of a solderless connector and the screw is then tightened. See Fig. 4–12.

The correct method for terminals for 10 AWG and smaller wires is shown in Fig. 4–13. Wrap the wire at least two-thirds (preferably three-quarters) of the way around the screw in a clockwise direction so that tightening the screw tends to close the loop rather than open it. Tighten the screw until it makes contact with the wire, and then tighten it about another half-turn to squeeze the wire a bit. As an alternative, tighten the screws to the pound-inches marked on the product, or to 12 pound-inches as recommended in Fig. 4–13. Never make the errors shown in Fig. 4–14. The insulation should end no more than ¼ inch from the screw at the most. Note that these two illustrations were made specifically for aluminum wire, but the principles are also correct for copper wire.

STEP 1: STRIP AND WRAP WIRE

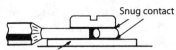

Contact plate on wiring device marked "CO/ALR"
STEP 2: TIGHTEN SCREW TO FULL CONTACT

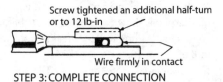

STEP 3: COMPLETE CONNECTION

Do not connect two wires under a single wraparound terminal screw even though it might appear logical in some situations. The *NEC* prohibits it in 110.14(A). Take those two wires and another short length of the same wire, and connect all three together

Fig. 4–13 Be sure wire is wrapped around terminal screw in clockwise fashion as in Step 1 so that tightening the screw tends to close the loop. Then complete Steps 2 and 3. *(UL Inc.)*

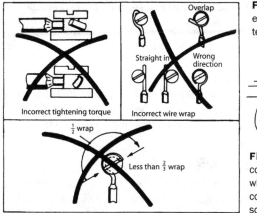

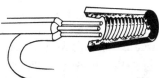

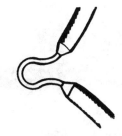

Fig. 4-14 Avoid these common errors when connecting wire to terminal screws. *(UL Inc.)*

Fig. 4-15 Use this method, commonly called a "pigtail" splice, when you would otherwise have to connect two wires under a terminal screw.

using a wire connector described later in this chapter. See Fig. 4–15. Then connect the remaining end of the short wire under the terminal screw. Most connectors of the kind shown in Fig. 4–12 are for one wire unless the connector is marked to indicate the number and size of wires that can be accommodated. The marking may be on the carton if the connector itself is too small for the marking.

What length should the bare wire be for connecting it under the terminal screw? Leave just enough bare wire to go around the screw, form a loop with a pair of long-nose pliers, slip it around the screw, close it with the pliers so that the loop is entirely under the screw, and then tighten the screw.

Push-in clamps Some 15-amp receptacles and switches have no terminal screws at all. Instead, there are internal terminal clamps that grasp a straight piece of wire pushed into them, forming an effective connection. Merely strip the end of the wire for half an inch or so (the proper length is usually shown on the device itself), and push the wire into the hole on the device. See Fig. 4–16. If an error is made, release the clamp by pushing a small screwdriver blade into another opening on the device. These push-in connections are acceptable only for copper or copper-clad aluminum

Fig. 4-16 Strip the wire, push it into opening on switch or receptacle, and the connection is made.

Fig. 4-17 In raceway wiring, a continuous wire may be connected to a terminal as shown here.

wire. They must not be used with all-aluminum wire. For receptacles these connections are limited to 14 AWG copper wire.

Connecting wires from raceway   In raceway wiring, sometimes a wire is pulled through one box and on to another, possibly through still another, and so on. If the wire merely runs through the box, pull it through without splice but leave a loop in case it is necessary to re-pull it in the future. If you intend to make a connection to the wire as it passes through the box, let a loop several inches long project out of the box. Strip away an inch or so of insulation, form a loop, and connect it under one terminal screw as shown in Fig. 4–17.

CONNECTORS FOR SPLICING WIRES

Connectors are used in splicing (joining) two or more pieces of wire together. It is important that whichever type of wire connector you use, be certain it is listed for the number and size of wires that are to be joined. The spliced wires must be electrically as good as an unbroken length of wire. The insulation of the splice must be as good as that on the original wires. Such a splice is accomplished by using properly installed insulated solderless connectors. Many people call these "Wire Nuts," which is the trade name of a particular manufacturer. When two or more ends of wire must be connected to each other, lay the wires together with their cut ends pointing in the same direction; insert the wires into the connector and turn it onto the wires, which twists them together as shown at the right in Fig. 4–18.

Connectors for joining two or more wire ends   One type of connector has a threaded metal insert molded into the insulating shell. Screw the connector onto the wires to be joined. The other kind has a removable metal insert. Slip the insert over the wires to be joined, tighten the screw of the insert, and then screw the insulating shell over the metal insert.

The spring-loaded connector shown in Fig. 4–19 is also popular. Inside its insulating shell there is a cone-shaped metal spring. Screw the connector over the wires to be joined. The insulating cover provides a good grip. When being screwed on, the coil spring temporarily unwraps. When released, it forms a very tight grip on the wire.

Fig. 4–18 Two types of solderless connectors for smaller wires, often called by the trade name "Wire Nut." The version with the setscrew maintains the integrity of the electrical connection even with the insulating cap removed. This makes it a good choice if the connection will be tested with diagnostic equipment while energized, such as with some motor connections.

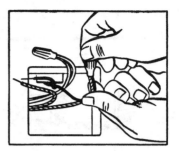

Fig. 4-19 This kind of connector contains a tapered coil spring inside the insulating cover.

In using these connectors, if one wire is much smaller than the others, let it project a bit beyond the heavier wires. If you have removed the right length of insulation from each wire, the insulating shell will cover all bare wires and no taping is necessary. Note that all types of these connectors are available in various sizes, depending on the number and size of the wires to be joined. Take care to observe the restrictions on allowable wire combinations that come with these connectors.

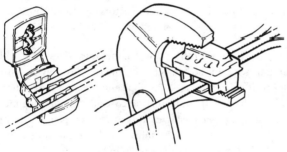

Fig. 4-20 "Clamshell" wire connector for taps or pigtails. For copper wire only.

For wire sizes 10 AWG and smaller, the insulated "clamshell" connector shown in Fig. 4-20 may be used. The wire insulation is used to position the wire in this type of connector, so do not strip the wire insulation before installing in the connector. A squeeze with parallel-jaw pliers installs the self-insulating connector.

Another wire connector, for 10 AWG and smaller wire, is the shell shown in Fig. 4-21. Unless the manufacturer's instructions direct otherwise, first twist the wires together, then slip the shell over them and crimp using a tool of the type shown. Then cut off the wire

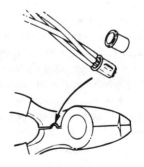

Fig. 4-21 Crimp-type wire connector, and pliers with special crimping die in handle. For copper wire only.

Fig. 4–22 For heavier wires, use metal connectors. The assembled connector and wires must be taped.

Fig. 4–23 Use a split-bolt connector when splicing a heavy wire to another continuous heavy wire. The assembled connector and wires must be taped.

ends that could puncture the insulation, and either tape or use formed plastic caps to insulate. These connectors are commonly used for joining bare grounding wires that do not require insulation.

For wires that are too large to be joined by the connectors described above, use heavy duty copper connectors of the style shown in Fig. 4–22 or Fig. 4–23.

Connectors for splicing to a continuous wire Sometimes a wire must be spliced to another continuous wire. In the heavier sizes, the simplest way is to use one of the split-bolt connectors shown in Fig. 4–23. Tape after making the connection. Some connectors are available with an insulating cover that can be snapped on after making the connection. For smaller sizes, it is usually simpler to cut the continuous wire to form two ends; the wire to be spliced in the connection makes the third wire. Then use a solderless connector as shown in Fig. 4–15.

Insulating splices Splicing devices such as solderless connectors are self-insulating. For other styles, insulating covers or boots are available that can be added after the splice is made. Some must be taped (See Figs. 4–22 and 4–23). For these splices, wrap all the exposed metal in self-vulcanizing rubber tape to make a thick, insulating blanket that fully equals that of the conductor insulation. Then wrap the connector and the ends of the wires with a high quality plastic tape, leaving none of the rubber showing. That will produce a safe result; but experienced electricians begin the process by wrapping the connector in a nonadhesive layer, such as varnished cambric tape. That prevents the rubber from fouling the threads on the split bolt, allowing it to be easily reusable if the connection needs to be remade. Begin and end the cambric with very short pieces of conventional electrical tape so it will stay in place as you wrap the rubber tape around it.

Chapter 5
CIRCUIT PROTECTION AND PLANNING

OVERCURRENT DEVICES SUCH AS FUSES and circuit breakers limit the amperage in any wire to the maximum that is permitted by the *National Electrical Code* (*NEC*). If more than the permissible maximum amperage is allowed to flow, the temperature of the wire goes up and the insulation may be damaged, leading to shortened life and accidental grounds that can become dangerous. If the overload is great enough, there is danger of fire. The maximum amperage established by the *NEC* as safe for any particular kind and size of wire under the conditions the wire is being used is called the ampacity. Table 4–1 lists the ampacity of common wires under standard conditions.

OVERCURRENT DEVICES
Any overcurrent device you use must have a rating in amperes not greater than the ampacity of the wire that it protects. However, since overcurrent devices, as a practical matter, need to be manufactured in standard sizes, the *NEC* allows the next higher standard size to be used in most cases. For example, if 12 AWG wire has a calculated ampacity, in a particular application, 20 amps, the circuit breaker or fuse that you use to protect the wire must have a rating not greater than 20 amps.

When two different sizes of wire are joined together (for example, when 8 AWG is used for mechanical strength in an overhead run to a building where it is joined to 14 AWG for the inside wiring), the overcurrent protection must be the right size for the smaller of the two wires. A fuse or circuit breaker of the correct size for the larger wire may be used at the starting point, provided another one of the proper size for the smaller wire is used where the wire is reduced in size. An example is when a wire such as 8 AWG, with an ampacity of 40, runs from one building to another where it feeds several 15-amp circuits (See Fig. 5–1).

What do you do when a fuse blows or a circuit breaker trips? Most people will say to install a new fuse or reset the breaker. That's correct—but first find out what caused the blown fuse or tripped breaker. Fuses and circuit breakers are the safety valves of electrical installations. Using substitutes or fuses too large for the size of

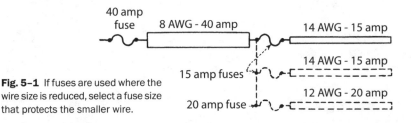

40 amp
fuse   8 AWG - 40 amp   14 AWG - 15 amp

14 AWG - 15 amp

15 amp fuses

12 AWG - 20 amp

20 amp fuse

Fig. 5–1 If fuses are used where the wire size is reduced, select a fuse size that protects the smaller wire.

wire can be dangerous because it could lead to a fire. Chapter 19 explains how to troubleshoot a blown fuse or tripped circuit breaker.

Fuses   A fuse is nothing more or less than a short piece of metal of a kind and size that will melt when more than a predetermined number of amperes flow through it. This metal link is enclosed in a convenient housing to prevent hot metal from spattering if the fuse blows and to permit easy replacement. A fuse rated at 15 amps is tested to carry 15 amps. When more than 15 amps flows through it, the wire inside the fuse melts (the fuse "blows"), which is the same as opening a switch or cutting the wire. The greater the overload, the quicker the fuse will blow.

Plug fuses   The common plug fuse, shown in Fig. 5–2, is made in ratings up to 30 amps. It is known as the Edison-base type because the base is the same as on ordinary lamps. These fuses are not permitted in new installations. They may be used only as replacements, and then only when there is no evidence of tampering or overfusing (using a fuse too large for the size of wire involved). For new installations, see the entry for Type S nontamperable fuses. Plug fuses are rated at 125 volts, but may be used on a system having a grounded neutral and no conductor over 150 volts to ground, so they could be used for a 240-volt load served from a 120/240 volt, three-wire system.

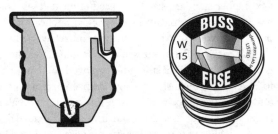

Fig. 5–2 Edison-base plug fuses are made only in ratings up to 30 amps. They are not permitted in new installations.

Time-delay fuses   When a motor is started, fuses often blow because a motor that draws only 6 amps while running may draw as much as 30 amps for a few seconds while starting. An ordinary fuse carries 80 percent of its rated amperage indefinitely, but blows very quickly if twice that amperage flows through it. However, wire that can safely carry 15 amps continuously—but might be damaged or even cause a fire if 30 amps flowed continuously—will not be damaged or cause a fire if 30 amps flows for a few seconds. The "time delay" type of fuse takes advantage of the ability of wire to carry a higher load momentarily. Commonly called by the trade name "Fusetron" although there are other brands, it looks like an ordinary fuse but is made differently inside. It blows just as quickly as an ordinary fuse on a small continuous overload or on a short circuit, but it will carry a big overload safely for a fraction of a minute. This type of fuse is convenient where motors are used because it prevents needless blowing of fuses and eliminates many service calls. Edison-base time-delay fuses are not permitted in new installations. They are permitted as replacements only, and then only if there is no evidence of overfusing or tampering. For new installations, see the entry for Type S nontamperable fuses.

Type S nontamperable fuses   The *NEC* requires the use of Type S fuses in all new installations that use fuses and for replacements if there is evidence of tampering or overfusing. Since all ratings of ordinary Edison-base plug fuses are interchangeable, nothing prevents someone from using, for example, a 25-amp

or 30-amp fuse to protect a 14 AWG wire, which must be protected at not over 15 amps. To prevent this unsafe practice, the nontamperable fuse was developed (See Fig. 5–3). The *NEC* calls this a "Type S" fuse, but it is commonly called by the trade name "Fustat." The fuse itself will not fit an ordinary

Fig. 5–3 A typical Type S nontamperable fuse (center, cross section at right, and its adapter to the left). Once an adapter has been screwed into a fuseholder, it cannot be removed. This prevents the use of fuses larger than originally intended.

fuseholder, so an adapter (shown at left in the illustration) must first be installed in the ordinary fuseholder; once installed, it cannot be removed. There are three sizes of adapters. As a safety measure, the 15-amp will accept only 15-amp or smaller fuses; the 20-amp will accept only 16-amp through 20-amp fuses; the 30-amp will accept 21-amp through 30-amp fuses. Type S fuses are presently made only in the time-delay type. They are rated at 125 volts, but may be used on a system having a grounded neutral and no conductor over 150 volts to ground, so they could be used for a 240-volt load served from a 120/240 volt, three-wire system.

A word of caution when using Type S fuses—when screwing the fuse into its holder, turn it some more after it appears to be tight. Under the shoulder of the fuse there is a spring—the fuse must be screwed in tightly enough to flatten the spring or the fuse will not "bottom" and you will have an open circuit just as if the fuse were blown.

Cartridge fuses   This type is made in all amperage ratings. Those rated at 60 amps or less are of the ferrule type shown in Fig. 5–4. Those rated at 70 amps or more have knife-blade terminals as shown in Fig. 5–5. The cartridge fuses of the type illustrated may be used in any circuit of up to 250 volts between conductors. Cartridge fuses are used for loads of 30 amps and higher such as mains, dryers, ranges, resistance heaters, and water heaters.

Fig. 5-4 Cartridge fuses rated 60 amps or less are of the ferrule type shown.

Fig. 5-5 Cartridge fuses rated more than 60 amps have knife-blade terminals as shown.

Circuit breakers   In new residential construction, circuit breakers are generally installed, with fuses appearing mostly in existing installations. A circuit breaker looks like a toggle switch. Figure 5-6 shows a single unit. Inside each breaker is a fairly simple mechanism which in case of overload trips the breaker and disconnects the load. If a breaker trips because of overload, in most brands you must force the handle beyond the OFF position, then return it to ON to reset it. On some brands, the

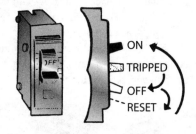

Fig. 5-6 A single circuit breaker (left) and method of resetting the breaker if it trips.

handle merely goes to the OFF position and is reset by returning it to the ON position.

A circuit breaker has a definite time delay. It will carry 80 percent of its rated load indefinitely, a small overload for a considerable time, and will trip quickly on a large overload. Nevertheless, it will carry temporary overloads long enough to permit motors to start.

FEEDERS AND BRANCH CIRCUITS

In larger homes, it may be advantageous to locate an additional panelboard near a concentration of required branch circuits, particularly in cases where that load is some distance from the service equipment. For example, if the kitchen were on the other side of the house from the service entrance, you might decide to place a panelboard in the basement under the kitchen. In doing so, you would supply this additional panelboard with a large circuit capable of supplying the calculated load of the individual branch circuits connected at this point. For example, you might install a 60-amp circuit breaker in the service panelboard, and then run a 6 AWG cable or wires in a raceway to the location of the smaller panelboard. This would be a practical application of the concept illustrated in Fig. 5-1. The official *NEC* term for the conductors extending between a service or other source of supply and the final branch-circuit overcurrent device is *feeder*.

The remote panelboard is often referred to as a "subpanel." Some time ago the *NEC* removed this term throughout the *NEC*, but it widely persists in the vernacular of the trade. Review Chapter 7, "Grounding Basics." In general, feeders must have their grounded conductors (white wires) insulated from their equipment grounding conductors (bare or green wires), as also noted in Chapter 8 under the topic "Bonding the busbar." This means that for a 240-volt feeder, there will usually be four wires, comprised of two ungrounded conductors (often black, or black and red), a neutral (white), and an equipment ground (a bare or green wire). If the feeder is wired with a metal raceway such as EMT that qualifies as a grounding conductor, the separate fourth wire may be omitted. Note that the judicious use of a feeder can reduce voltage drop, because the portion of any total circuit run from the service to the outlet that is taken up in a feeder occurs over a larger conductor with less resistance per unit length. The remote panelboard may also be located in a more convenient location. However, the requirement for appropriate workspace about the remote panel is the same as that for the service panelboard. Review the topic "Service equipment location" in Chapter 8 for more details.

CALCULATING BRANCH CIRCUIT NEEDS

If all the lights and appliances in a home or on an entire farm were protected by a single fuse or circuit breaker, the entire establishment would be in darkness when that breaker tripped or the fuse blew. Also, all the wires would have to be very large to match the ampere rating of that breaker or fuse, which would make a clumsy and expensive installation. Therefore the different outlets in an installation are separated into smaller groups known as branch circuits. Inexpensive 14 AWG or 12 AWG wire is used for most of the wiring, protected by 15-amp or 20-amp breakers or fuses. When one of these breakers trips or a fuse blows, only the outlets on that circuit are dead; those on other circuits are still live.

Continuous loads   See the discussion under "Water heaters" on page 170 in Chapter 14, "Appliances."

Outlets per circuit   In residential work, the *NEC* does not limit the number of lighting outlets placed on one circuit. But if you put on too many, you will probably have trouble with breakers tripping or fuses blowing. In most cases, it is best to connect fewer than a dozen outlets on one circuit even if more are permitted.

Note: An outlet is any point where electric power is actually used. Each fixture, even if it has five lamps, is considered one outlet. Each receptacle outlet, even the duplex type, is one outlet. Switches are not outlets since they use no electric power, but merely control its use. But in estimating the cost of an installation, switches are included in counting the outlets to arrive at a total cost on the "per outlet" basis, though this is not in accordance with the *NEC* definition. Many electricians quote on a "per opening" basis to avoid this terminology problem.

Kinds of circuits   The circuits used in homes can be divided into the following three general types: *lighting circuits*—primarily for lighting and serving permanently installed lighting fixtures, as well as receptacle outlets into which you plug lamps, radios, televisions, clocks, and similar 120-volt loads (but not kitchen appliances); *small appliance circuits*—receptacles in kitchen, dining room, etc. for such items as coffeemaker, toaster, electric fry pan; *individual appliance circuits*—each serving a single appliance such as a range, water heater, clothes washer or dryer; circuits may be either 120-volt or 240-volt.

In addition, there may be a need to add a circuit for a permanently connected motor that is not part of an appliance, such as might be used with certain shop tools. (Motor types and installation are discussed in Chapter 16.)

Lighting circuits   The NEC requires enough lighting circuits to provide 3 VA of power for every square foot of floor space in the house. Since a circuit wired with 14 AWG wire and protected by 15-amp overcurrent protection provides 1,800 VA (15 × 120 = 1,800), each circuit is enough for 600 square feet (1,800 ÷ 3 = 600). This is the NEC *minimum*. The NEC concerns itself with safety only. To allow for convenience and usefulness over time, it is wise to provide one circuit for each 500 square feet of space.

What is included in the square footage of a house when calculating the number of lighting circuits needed? Open porches and garages, even if attached to the house, are not included in the total. But unfinished or unused spaces that are adaptable for future use must be included according to the NEC. Run a special circuit to such an area, terminate it in a single outlet if you wish, and later branch off from that outlet to the additional outlets that you want when you finish off the space. If the space is large, run two circuits.

To arrive at the NEC definition of the total number of square feet for determining the number of circuits, multiply the length of the house by its width. If its outside dimensions are 30 × 45 feet, then one floor represents 1,350 square feet (30 × 45 = 1,350). If the basement is finished or can be finished into usable space, add its area of, for example, 24 × 30 or 720 square feet to the first-floor area of 1,350 square feet for a total of 2,070 square feet. If an upper floor has unfinished space that can later be finished into a bedroom, add its area.

Then divide the total area by 600 to arrive at the minimum number of lighting circuits the NEC requires, or divide by 500 for a more adequate installation. In Table 5–1, minimum and recommended numbers of lighting circuits are calculated for some typical square footages.

Special small-appliance circuits   The circuits already discussed are for lighting, including floor and table lamps, and items such as radios, TVs, and vacuum cleaners. These circuits do not permit proper operation of larger kitchen and similar appliances that consume much more power. NEC 210.52(B) and 210.11(C)(1) require two special small-appliance circuits to serve only small-appliance outlets, including refrigeration equipment, in the part of the house occupied by kitchen, pantry, breakfast room, and dining room. Both circuits must extend to the kitchen counters; the other rooms

Table 5–1 NUMBER OF LIGHTING CIRCUITS		
USABLE AREA	CODE MINIMUM	RECOMMENDED
1,200 sq ft	2	3
1,600 sq ft	3	4
2,000 sq ft	4	4
2,400 sq ft	4	5
2,800 sq ft	5	6
3,200 sq ft	6	7

may be served by either or both of them. The circuits must be wired with 12 AWG wire protected by 20-amp circuit breakers or fuses. No lighting outlets may be connected to these circuits (with two exceptions: a combination receptacle/ support for an electric clock in kitchen or dining area, and a receptacle serving the ignition system, oven light, and/or timer for a gas-fired cooking appliance). Either 15-amp or 20-amp 125-volt rated receptacles may be installed on these 20-amp circuits.

Each such circuit has a capacity of 2,400 VA (20 × 120 = 2,400), which is not too much considering that toasters, irons, and similar appliances often require 1,000 watts, and appliances such as roasters and toaster ovens may consume over 1,500 watts. These two circuits can be merged into one multiwire circuit, which is discussed on pages 118–119.

At least one separate 20-amp circuit must be run to laundry appliances, and another one to all the bathroom receptacles (or multiple loads in just one bathroom). A separate 15-amp branch circuit is permitted to serve refrigeration equipment.

Individual circuits   It is customary to provide a separate circuit for each of the following (and required for central heating equipment):

- Self-contained range
- Separate oven, or counter-mounted cooking unit
- Water heater
- Clothes washer
- Clothes dryer
- Waste disposer
- Dishwasher
- Motor (and fan) on oil-burning furnace
- Motor on blower in gas furnace
- Water pump
- Permanently connected appliances rated at more than 1,000 watts (Example: a bathroom heater)
- Permanently connected motors rated more than ⅛ hp

Note that on general-purpose branch circuits (those supplying two or more receptacles or outlets for lighting or appliances), the maximum load of utilization equipment fastened in place (other than fixtures), such as exhaust fans, garage door openers, etc. cannot exceed 50 percent of the branch circuit rating. This is often the deciding factor in whether or not to run an individual circuit. Appliance and motor circuits may be either 120-volt or 240-volt depending on the particular item installed. These circuits are discussed in more detail in Chapter 14, "Appliances." See page 220 for instructions on *NEC* requirements for labeling every circuit at the panel-board to assist in troubleshooting.

Motor circuits   Separate circuits are recommended for individual motors over

1/8 hp that are not part of an appliance. Such motors might be either directly connected (example: sump pump) or belt-driven (example: bench grinder). Determine the motor's amperage from Table 5-2, adapted from *NEC* Table 430.248. See also the discussion on pages 181–182.

Total number of circuits You must decide for yourself how many circuits you will need. Provide circuit breakers or fuses for specific appliances even if you do not intend to install such appliances

Table 5-2	DETERMINING CORRECT AMPERAGE FOR MOTORS	
MOTOR	120 VOLTS	240 VOLTS
¼ hp	6 amps	3 amps
⅓ hp	7 amps	3½ amps
½ hp	10 amps	5 amps
¾ hp	14 amps	7 amps
1 hp	16 amps	8 amps
1½ hp	20 amps	10 amps
2 hp	24 amps	12 amps
3 hp	34 amps	17 amps
5 hp	56 amps	28 amps

until later. Even then, provide for some spare circuits. For a new wiring installation, follow the suggestions in Table 5–3 for calculating the total number of 120-volt and 240-volt circuits needed based on the square footage of usable area.

Arc-fault circuit interrupters (AFCI) The US Consumer Product Safety Commission estimates that annually there are more than 40,000 fires in residential

Table 5-3	SUGGESTED NUMBER OF 120-VOLT AND 240-VOLT CIRCUITS					
Square feet of area in house ▶ 1,200	**1,600**	**2,000**	**2,400**	**2,800**	**3,200**	
CALCULATE 120-VOLT CIRCUITS General-purpose circuits (one per 500 sq. ft.)	3	4	4	5	6	7
Spare general-purpose circuits	1	1	2	2	2	2
Small appliance	2	2	2	2	2	2
Special laundry	1	1	1	1	1	1
Bathroom receptacles	1	1	1	1	1	1
Individual appliance circuits:						
Oil or gas burner	1	1	1	1	1	1
Blower on furnace	1	1	1	1	1	1
Bathroom heater, room air con- ditioner, workshop motor, etc.	1 each	1 each	1 each	1 each	1 each	1 each
Spares	1	1	2	2	2	2
Fill in total 120-volt circuits ▶						
CALCULATE 240-VOLT CIRCUITS *Individual appliance circuits:*						
Range	1	1	1	1	1	1
Water heater	1	1	1	1	1	1
Clothes dryer	1	1	1	1	1	1
Water pump, other fixed in place	1 each	1 each	1 each	1 each	1 each	1 each
Spares	1	1	1	1	1	1
Fill in total 240-volt circuits ▶						

occupancies, resulting in 250 lives lost and $1 billion in financial loss. It has been estimated that 40 percent of these fires are caused by arcing faults—the unwanted flow of electricity through an insulating medium (such as air). An arc can generate enough heat to be a source of ignition, but not enough current to trip the standard circuit breaker.

Just as there are circuit breakers incorporating ground-fault protection for the purpose of preventing people from receiving dangerous shocks (See pages 74–75), there are now circuit breakers incorporating arc-fault circuit interruption in addition to the normal overcurrent functions, and some that incorporate both GFCI and AFCI in the same breaker frame. Arcing faults exhibit a characteristic current and voltage pattern that can be detected by the electronics in the AFCI, which then interrupts the circuit.

Arc-fault circuit-interrupters were originally required for all circuits supplying outlets, for both receptacles and otherwise, in dwelling unit bedrooms. Subsequent editions dramatically expanded the scope of this requirement, which now includes all outlets and devices (includes switches) in kitchens, family, dining, and living rooms, as well as those in parlors, libraries, dens, sunrooms, recreation rooms, and closets. In case you are in doubt, the rule adds "similar rooms or areas" for good measure. The best option, therefore, is to use AFCI protection for all indoor electrical openings in finished areas except bathrooms. These rooms have GFCI protection in place for all receptacle outlets. Although GFCI protection is primarily shock protection, since arcing faults usually involve a grounded surface, and since such currents to ground will immediately trip a GFCI protective device, bathrooms still have substantial protection.

Most (but not all) AFCI circuit breakers cannot be used on multiwire branch circuits (See Fig. 10–15 and associated text), although this is slowly changing. This means that if existing wiring involves this relatively common wiring arrangement, you should discuss with the local inspector the extent to which this rule will be enforced on existing circuits.

Multiwire circuits are very common, and many homes wired before the early 1960s have at least some branch circuits that have no equipment grounding conductor. As the industry gets up to speed on the full ramifications of these AFCI requirements, look for a steady evolution in the ways these devices can be installed. At least one manufacturer now has two-pole AFCI circuit breakers that fit in both the usual clip-on connections, and a different line that fits in "bolt-on" panelboards.

As of the 2011 edition, the NEC requires that all wiring "modified, replaced, or extended" in the covered areas have AFCI protection, either through the usual AFCI circuit-breaker approach, or through an AFCI receptacle positioned as the first one in the circuit. Such devices have been listed, and are now in production (See Fig. 5–11). In addition, as of January 1, 2014, AFCI protection is required at or ahead of any receptacles being replaced in locations where the applicable edition of the NEC requires AFCI protection, resulting in an even wider market in existing dwelling units.

The 2014 *NEC* includes six acceptable methods for meeting the AFCI requirement, and all but the first employ one of the new "outlet branch-circuit type AFCI" devices (the formal terminology for what is generally known as an AFCI receptacle, also called an " AFCI"). The first method, and the only one previously on the market, is a "combination-type" AFCI circuit breaker that protects the entire branch circuit. Given its downstream location in a circuit, an OBC cannot protect against a parallel event on its line side, and the remaining five options address this limitation. Refer to *NEC* 210.12(A) for the exact rules covering these options, because they are complicated and in some cases involve protective devices that may have limited availability in the market.

Fig. 5–7 An outlet-branch-circuit (OBC) type AFCI receptacle capable of responding to both parallel and series arcs on it load side, and series failures on its supply side. (*Leviton*)

Chapter 6
CIRCUIT DIAGRAMS

BEFORE YOU CAN WIRE A building, you must learn how switches, receptacles, sockets, and other devices are properly connected to each other with wire to make a complete electrical system called a circuit—the path along which current flows.

GROUNDED WIRES

In residential and farm wiring one of the current-carrying wires is grounded, which means it is connected to an underground metal water pipe system and to a driven ground rod. Many people refer to the grounded wire as the neutral wire. Per revisions in the 2008 *NEC,* a wire is defined as neutral if it is connected to the neutral point of an electrical system, even if all the associated ungrounded conductors are not present. Virtually all grounded conductors within the scope of this book are neutrals under the current *NEC* definitions. (Chapter 7 has a full discussion of grounding, and the Index at the back of the book lists specific grounding topics.) Ungrounded wires are known as "live" or "hot" or "energized" wires, which means they have a voltage, in house wiring usually 120 volts, above zero (ground potential—the earth is assumed to be at zero potential).

It is essential to remember these five points about grounding:

■ The grounded wire is white in color. Two recognized alternatives to white for the grounded conductor are: gray (*Caution:* white is preferred, and for many years gray theoretically could have been and occasionally was used for ungrounded conductors); or three white stripes on any color insulation except green.

■ The grounded wire must run direct to every 120-volt device or outlet to be operated (never to anything operating only at 240 volts).

■ The grounded wire is never fused or protected by a circuit breaker.

■ The grounded wire is never switched or interrupted in any other way.

■ White wire must never be used except as a grounded wire. Other wires are usually black but may be some other color, but not white or green.

The white wire must run to every 120-volt device or outlet other than a switch; the other wire to the device is usually black but may be some other color, but not white or green. White wire can be used for ungrounded purposes when it is part of a cable assembly brought to a switch or for a 240-volt load, but even here it must be reidentified as other than white at all visible points and at terminations. Examine a socket or similar device (used on 120-volt circuits) carefully and you will find that one of its two terminals is a natural brass color, and the other is silver-colored (usually nickel-plated) and called "white" by the *NEC*. The white wire must always run to the white terminal. In the case of sockets, the white terminal is always connected to the screw shell and never to the center contact of the socket. Switches never have white terminals for their power connections.

WIRING DIAGRAMS

Wiring diagrams are used in planning electrical installations. Diagrams are useful for planning safe and efficient installations, and for clarifying details such as the wiring of three-way and four-way switches.

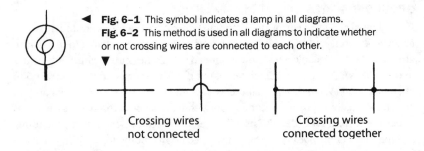

◀ Fig. 6–1 This symbol indicates a lamp in all diagrams.
Fig. 6–2 This method is used in all diagrams to indicate whether or not crossing wires are connected to each other.

▼

Crossing wires
not connected

Crossing wires
connected together

Diagram symbols   For any circuit, two wires must run from the starting point to each outlet serving electrical equipment (lamp, motor, etc.). Imagine that the current flows out over the black wire to the equipment, through the equipment, and back to the starting point over the white wire. In this book, the starting point is called the source, and a lamp symbol (Fig. 6–1) is used to indicate the equipment to be operated. Figure 6–2 illustrates the method used to indicate whether or not crossing wires are connected to each other. In all diagrams wires are indicated like this:

white wire   black wire   black wire between a switch and
                                                            the outlet which it controls

The fact that one line is heavier than the other does not mean that one wire is larger than the other; both are the same size.

Adding switches   The simplest possible diagram is that of Fig. 6–3, a lamp that is always on with no way of turning it off. Such a circuit is of little value, so you should add a switch. Figure 6–4 shows how you simply make the black wire

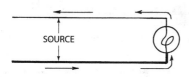

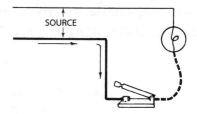

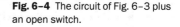

Fig. 6-3 The simplest circuit—a lamp always on.

Fig. 6-4 The circuit of Fig. 6-3 plus an open switch.

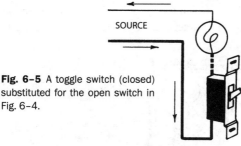

Fig. 6-5 A toggle switch (closed) substituted for the open switch in Fig. 6-4.

detour to a switch. An unenclosed switch is shown so you can see how it operates: opening (turning off) the switch is the same as cutting a wire. Of course, unenclosed switches are unsafe and not permitted. Figure 6-5 shows the same circuit but with an enclosed flush toggle switch which serves the same purpose, but is much neater, more convenient, and safer to use.

Wiring in parallel   If several lamps are used, do not make the mistake of wiring them *in series* as in Fig. 6-6 because that is impractical for most purposes. If one lamp burns out or is removed from its socket, it is the same as opening a switch in the circuit—all the lamps go out. Instead, wire them *in parallel* as in Fig. 6-7. The

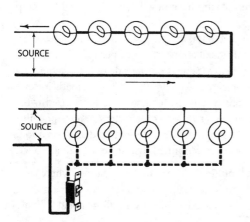

Fig. 6-6 Style of wiring known as "series" wiring. It is impractical except for very special purposes.

Fig. 6-7 Style of wiring known as "parallel" wiring. It is used for all ordinary purposes.

white wire goes to each lamp, from the first to the second and to the third and so on; the black wire also runs to each lamp in similar fashion. Each lamp will light even if one or more of the others is removed. The switch shown cuts off the current to all the lamps at the same time. The wiring is simple: white wire from SOURCE to each socket or outlet; black wire from SOURCE to the switch, and from the switch another black wire to each of the sockets. This is the way a fixture with five sockets is wired.

If each lamp is to be controlled by a separate switch, use the Fig. 6–8 diagram. The white wire always runs from SOURCE to each lamp; the black from SOURCE to each switch, and from each switch a black wire runs to the lamp that it controls. If you trace the current from the SOURCE over the black wire through each switch, one at a time, along the black wire to the lamp, and back from the lamp to the SOURCE, you will find that each lamp can be independently controlled.

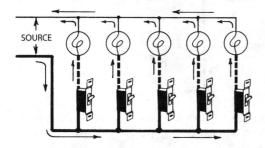

Fig. 6–8 This diagram shows the same parallel wiring as Fig. 6–7, but with each lamp controlled by a separate switch. Note that the white wire runs directly from SOURCE to each lamp.

Wiring receptacle outlets The wiring of plug-in receptacle outlets is simple. If there is only one receptacle, run the white wire from the SOURCE to the side of the receptacle that has white (silver-colored) terminal screws, and run the black wire to the other side, as in Fig. 6–9. If there are several receptacles, as in Fig. 6–10, run the white wire to one side of the first receptacle, from there to the second receptacle, and so on; do the same with the black wire. Figure 6–10 is the same parallel wiring diagram as Fig. 6–7 except that receptacles have now been substituted for lamps and the switch has been omitted.

Where the two wires shown in Fig. 6–10 are part of a three-wire circuit (See pages 118–119 for three-wire circuit installation), it is not permitted to feed the white conductor through by using the two screws on the side of the receptacle, as shown at *A*, because removing this receptacle could place 120-volt loads in series on 240 volts, with possibly disastrous results. Instead, a splice is required with a pigtail going to the receptacle as shown at *B*. Many inspectors require pigtailing at all feed-through receptacles even though the devices are tested and listed to be connected as shown at *A*. (See Fig. 4–15 for another illustration of a pigtail splice.)

Wiring outlets controlled by pull chains There are few locations where a pull chain will be used, since lighting in attics and under-floor spaces used for storage or for equipment requiring servicing must be controlled by a wall switch at the entry

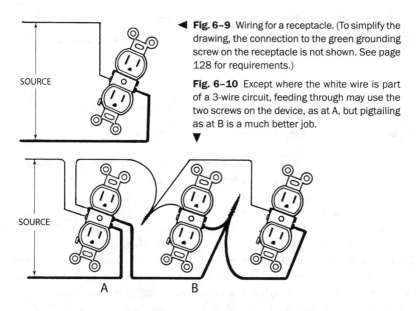

◀ **Fig. 6–9** Wiring for a receptacle. (To simplify the drawing, the connection to the green grounding screw on the receptacle is not shown. See page 128 for requirements.)

Fig. 6–10 Except where the white wire is part of a 3-wire circuit, feeding through may use the two screws on the device, as at A, but pigtailing as at B is a much better job.

▼

A   B

to the space. To wire for a pull chain, substitute a pull-chain socket (shown at *F* in Fig. 9–4) for the receptacle in Fig. 6–9.

Combining several diagrams   When several groups of outlets are to be wired, the two separate wires from each group can be run back to a common starting point where the wires enter the house, as shown in Fig. 6–11 where the diagrams of Figs. 6–5, 6–7, and 6–8 have been combined. This method requires a considerable amount of material because the wires must be much longer than necessary. It

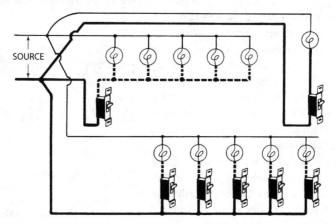

Fig. 6–11 The outlets of Figs. 6–5, 6–7, and 6–8 have been combined into a single group of outlets. In other words, here they form one circuit. This still uses a great deal of material. To reduce the amount required, the circuit can be arranged as the next diagram shows.

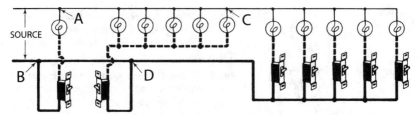

Fig. 6-12 The diagram of Fig. 6-11 rearranged to use less material.

is more efficient just to wire the first group, then run wires from the first group to the second, and from the second to the third, as in Fig. 6-12, which shows the same three groups but using less material.

The white wire may be extended at any point to the next outlet. The black wire likewise may be extended at any point, provided that it can be traced all the way back to the SOURCE without interruption by a switch. In other words, in these diagrams a black wire can be extended from any black wire indicated by the solid heavier lines, but not from the black wire between a switch and an outlet indicated by a heavy broken line. Thus, in Fig. 6-12, A and B are the starting points for the second group, and C and D are the starting points for the third group of lights.

Three-way switches The switch used for controlling a light from one point is known as a "single pole" switch. "Three way" switches allow you to control a light from two points (not three, despite the name). With three-way switches you can turn a hall light on or off from upstairs and downstairs, or a garage light from house and garage, or a yard light from house and barn. Such switches have three different terminals for wires. Their internal construction is similar to Fig. 6-13. In one position of the handle, terminal A is connected inside the switch to terminal C; in the other position, terminal A is connected to terminal B. Usually, the common terminal A is identified by being a darker color than the other terminals which are natural brass.

Study the diagram of Fig. 6-14 on page 59 in which the handles of both switches are down. The current can be traced from SOURCE over the black wire through *Switch 1*, through terminal A, out through terminal B and up to terminal C of *Switch 2*, but there it stops. The light is off.

Next see Fig. 6-15 where the handles of both switches are up. The current can be traced as before from SOURCE over the black wire through *Switch 1*, through terminal A, but this time out through terminal C, and from there up to terminal B of *Switch 2*. There it stops and the light is off.

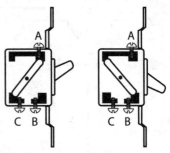

Fig. 6-13 This shows what happens inside a 3-way switch when the handle is thrown from one position to another. The current enters by the common terminal A, and leaves by either C or B depending on the position of the handle.

Now examine the diagrams of Figs. 6–16 and 6–17. In both diagrams the handle of one switch is up, the other down. In each case the current can be traced from SOURCE over the black wire, through both switches, through the lamp, and back to SOURCE. The light in both cases is on. In the case of either Fig. 6–16 or Fig. 6–17, throwing either switch to the opposite position changes the diagram back to either Fig. 6–14 or Fig. 6–15, and the light is off. In other words, the light can be turned on and off from either switch.

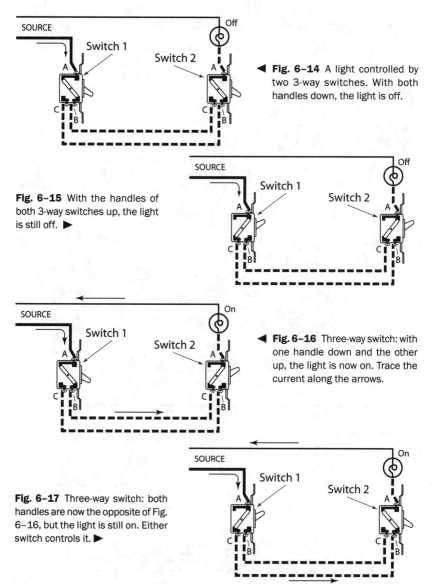

Fig. 6–14 A light controlled by two 3-way switches. With both handles down, the light is off.

Fig. 6–15 With the handles of both 3-way switches up, the light is still off. ▶

Fig. 6–16 Three-way switch: with one handle down and the other up, the light is now on. Trace the current along the arrows.

Fig. 6–17 Three-way switch: both handles are now the opposite of Fig. 6–16, but the light is still on. Either switch controls it. ▶

The wiring of three-way switches is simple, as these diagrams show. Run the white wire from SOURCE as usual to the light to be controlled. Run the black wire from SOURCE to the common or marked terminal of the first three-way switch. Run a black wire from the common or marked terminal of the second three-way switch to the light. That leaves two unused terminals on each switch; run two black wires from the terminals of the first switch to the two terminals of the second switch.

It makes no difference whether you run a wire from *B* of the first switch to *C* of the second as shown, or from *B* of the first to *B* of the second. The wires that start at one switch and end at another are called runners, travelers, or jockey legs. Because

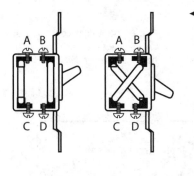

◀ **Fig. 6–18** Internal connections of one style of 4-way switch with the handle in the up and the down positions. The terminals are labeled *A, B, C,* and *D.*

Fig. 6–19 Four-way switches are usually connected as shown here. With the 4-way switch handles in the positions shown in the upper diagram, the light is off; in the lower diagram, the light is on. Two 4-way switches are shown in the circuit, but any number may be installed depending on need. Refer to Fig. 6–18 to identify the terminals. ▼

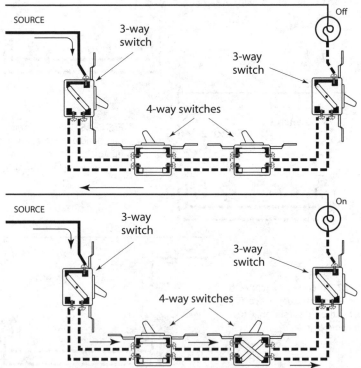

the load could be energized in either position of a three-way switch, these switches do not indicate ON and OFF positions.

Four-way switches   To control a light from more than two points, use two three-way switches, one nearest the SOURCE and the other nearest the light, and four-way switches at all remaining points in between. Four-way switches can be identified by

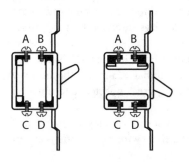

◀ **Fig. 6–20** Internal connections of the other style of 4-way switch with the handle in the up and the down positions. Terminals are labeled *A*, *B*, *C*, and *D*.

Fig. 6–21 Some brands of 4-way switches are connected as shown. With the 4-way switch handles in the positions shown in the upper diagram, the light is off; in the lower diagram, the light is on. Refer to Fig. 6–20 to identify terminals. ▼

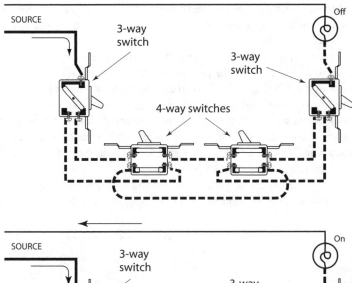

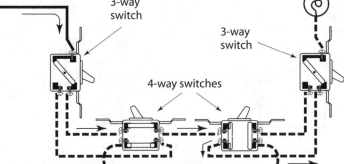

the fact that they have four terminal screws and do not have ON-OFF markings on the handles. Figures 6–18 through 6–21 show two types of four-way switches and the wiring diagrams for circuits using each type. Depending on the manufacturer, the internal connections of four-way switches may be as shown in Fig. 6–18 or as in Fig. 6–20. In each of these figures, the four terminals are labeled *A, B, C,* and *D*.

The four-way switch in Fig. 6–18 is shown in a circuit in Fig. 6–19. The two jockey legs from one three-way switch are connected to terminals *A* and *B* of the four-way switch; from the other three-way, the jockey legs are connected to terminals *C* and *D* of the four-way.

The four-way switch in Fig. 6–20 is shown in a circuit in Fig. 6–21. The jockey legs from one three-way switch are connected to terminals *A* and *D* of the four-way switch, and the jockey legs from the other three-way are connected to terminals *B* and *C* of the four-way switch.

If you are not sure which type of four-way switch you have, try one diagram; if it does not work, try the other. You can do no harm by wrong connections except that the circuit will not work if you have the wrong connection.

Pilot lights   When a light can't be seen from the switch that controls it, the light is often left on unnecessarily. The problem often occurs with basement and attic lights in homes and haymow lights in barns. Install a pilot light (a small, low-wattage lamp) near the switch so that both lights are turned on and off at the same time. The pilot light serves as a reminder that the unseen light is on. The wiring is shown in Fig. 6–22. First look at the diagram while disregarding the wires shown in dotted lines as well as the pilot light itself. It is then the same as Fig. 6–5. Add the wires shown in dotted lines and it becomes the same as the diagram of Fig. 6–7 except there are two lamps instead of five. The white wire runs to both lamps; the black wire from the SOURCE runs to the switch as usual; the black wire from the switch runs to both lamps, and the diagram is finished.

A combination switch and pilot light in one device is shown in Fig. 6–23. The diagram is correct only for some brands; exact diagrams usually come with each

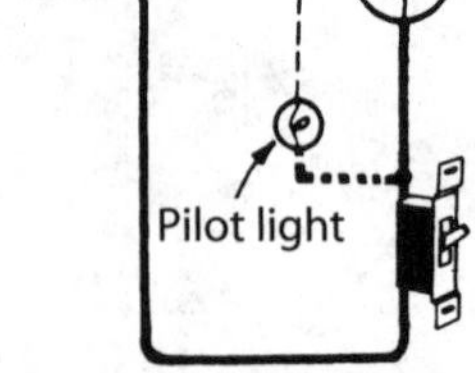

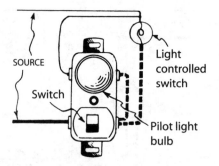

Fig. 6–22 The basic circuit of a pilot light.

Fig. 6–23 The wiring scheme of a combination pilot light and switch. Another brand might be wired somewhat differently.

device. There are also switches having a small pilot light in the handle. Some have the same lit handle except that it is on while the switch is in the OFF position to aid in finding the switch in the dark.

Other diagrams and combinations   You will have little difficulty making a diagram for any desired combination of outlets and switches, whether single-pole or three-way, if you remember the principles covered in this chapter. To make any diagram, first locate each outlet where desired and run the white wire (light line like this ────) from SOURCE to every outlet. Then run a black wire (heavy line like this ━━━━) from SOURCE to each outlet that is not controlled by a switch. Next run black wire (heavy broken line like this ▪ ▪ ▪ ▪) from each outlet controlled by a switch to the switch that controls it. If three-way switches are involved, run additional black wires between the switches that control the outlet. Finally, from each switch run black wire (heavy line like this ━━━━) back to SOURCE. Having done that, you may have to rearrange some of the wires to reduce the amount of wire used, but this will not affect the proper operation of the hook-up.

Anticipation of Occupancy Sensor Switches   The *NEC* is anticipating the increasing widespread utilization of occupancy sensing switches and the likely eventual requirement that they function through a connection to the actual grounded conductor, and not by leaking small currents into the equipment grounding system. To this end, if the wiring system does not provide means for readily bringing a properly connected white wire to the switch, then one needs to be installed for that eventual purpose when the switch is originally wired. The *NEC* defines an acceptable basis for future ready access to add a grounded conductor as being a raceway wiring method with enough room for another conductor to be pulled in, or easy access for rewiring with different cable, as in common commercial applications where the back side of the wall is left open, or where the surface treatment of the wall ends above a suspended ceiling or below a raised floor on the same floor level. These options are not commonly available in usual residential construction.

The *NEC* also waives the requirement in instances where the application is such that an occupancy switch would not or could not be installed. The neutral is not required where the switch does not serve a habitable room other than a bathroom; a frequently cited example of this is a door jamb switch controlling a closet light. If there are multiple switch locations that "see" the entire lit space, then the neutral need be brought to only one of those locations. Occupancy sensors will never be listed for the control of a receptacle load, because such loads are unknowable; and, therefore, such switch locations need not be wired with neutrals. Some locations use snap switches in series with occupancy sensors so the lights are sure to go off when the room is unoccupied, and also forced off by a switch in order to guarantee darkness, such as for a projected presentation. Such switch locations clearly do not require a neutral. Note that a neutral is required, however, for a switch location outside but that still controls the lit area. Although an occupancy sensor would never replace such a switch, some high-end dimmers require a neutral, and therefore the Code-Making Panel elected to leave the requirement in place for these locations.

Chapter 7
GROUNDING FOR SAFETY

THE PRACTICAL USE OF ELECTRIC POWER involves the generation of electromotive force, often abbreviated as "emf." Since the amount of electrical force or pressure is measured in volts, a very common term for this is voltage. In ordinary electrical systems, voltage is maintained between conductors. As of the 2008 *NEC*, the term *"ground"* is equated to *"earth."*

All *NEC*-recognized electrical systems maintain a direct connection between their metal enclosures (conduits, metal cable assemblies, switch and panel enclosures, etc.) and the earth through one or more buried *grounding electrodes* and the *grounding electrode conductors* that connect one to the other. The idea is to limit the voltage to ground on any enclosure so it will be safe to touch, even under fault conditions. The mechanism that unites all conductive electrical enclosures and wiring methods into a single conductive path is the *equipment grounding* system, and the conductor associated with any given circuit is the *equipment grounding conductor* for that circuit. The principal focus of equipment grounding is shock protection.

Grounded systems have safety benefits If one of the circuit conductors is directly "connected to ground or to a conductive body that extends the ground connection" (*NEC* definition), then there will be a definite voltage to ground on the other circuit conductors. A system so arranged has a *grounded conductor* and is referred to as a *grounded system*. All systems covered in this book are grounded systems. Grounded systems have certain safety features that make them crucial for wiring for general use. These features can be divided into two general topics.

System performance The grounded circuit connection tends to stabilize the voltage to ground against transient high voltage events. For example, suppose a 2,400-volt line accidentally falls across your 120/240-volt service during a storm. If the system is not grounded, you can be subject to deadly 2,400-volt shocks, and wiring and appliances will be ruined. If the system is properly grounded, the highest voltage of a shock will be much more than 120 volts (the voltage to ground), but very much less than 2,400 volts. Lightning striking on or even near a high voltage line can cause great damage to your wiring and your appliances, and it can cause fire and injuries. Proper grounding throughout the system greatly reduces the danger.

Fault clearance and shock protection   Recall that all metal electrical enclosures are connected to the earth, for both grounded and ungrounded systems. The principal benefit of a grounded system is that one of the circuit conductors shares the same connection. In the event of an insulation failure that puts a hazardous voltage on electrical equipment, that electrical connection at the point of failure becomes, in effect, a short circuit between the grounded circuit conductor and the ungrounded conductor involved in the fault. The result will be that the fuse will blow or the circuit breaker will trip immediately, removing the shock hazard. See the section "How grounding promotes safety" in this chapter for practical examples of this principle at work.

Because grounding is so important, it is discussed repeatedly throughout this book. Be certain that you understand it thoroughly. The *National Electrical Code (NEC)* rules for grounding are extensive and sometimes seem ambiguous. But for installations in homes and farm buildings—the only installations covered in this book—the rules are relatively simple. This chapter discusses the basic principles of grounding. The practical details, including installing ground rods and ground clamps and selecting the correct ground wire size, are covered in the next chapter on installing the service entrance.

HOW DANGEROUS ARE SHOCKS?

Most people think it is a high voltage that causes fatal shocks. This is not necessarily so. The amount of current flowing through the body determines the effect of a shock. A milliampere (mA) is one-thousandth of an ampere (0.001 amp). A current of 1 mA through the body is just barely perceptible. Currents from 1 to 8 mA cause mild to strong surprise. Currents from 8 to 15 mA are unpleasant, but usually the victim is able to let go and get free of the object that is causing the shock. Currents over 15 mA are likely to lead to "muscular freeze," which prevents the victim from letting go and often leads to death. Currents over 75 mA are almost always fatal; much depends on the individual involved.

The higher the voltage, the higher the number of milliamperes that would flow through the body under any given set of circumstances. A shock from a relatively high voltage while the victim is standing on a completely dry surface will result in fewer milliamperes than a shock from a much lower voltage to someone standing in water. Deaths have been caused by shock from circuits considerably below 120 volts, while someone standing on a dry surface could survive shock from a circuit of 600 volts and more. How to help the victim of a shock is discussed on page 7.

GROUNDING BASICS

This section, together with the introductory passages in this chapter, defines grounding terminology and describes the functions of the wires and equipment in your installation. Italics are used to indicate words for which definitions are provided and to help differentiate among similar-looking terms.

Grounding electrode The grounding electrodes that are present provide the electrical interface between a wiring system and the earth. Actually, the *NEC* refers to the "grounding electrode system"—which must be comprised of whatever qualified electrodes of the types listed in *NEC* 250.52 are present. These systems are comprised of one or more "grounding electrodes," which may include the familiar underground metal water piping supplemented by one or more driven metal rods or pipes as illustrated at the bottom of Fig. 8–6. The ground rod or pipe may be the only electrode if no water pipe or other qualifying electrode is present. Refer to the discussion on page 92 for more information.

Bonding (or bonded) Bonding simply means connecting so as to establish electrical continuity and conductivity. The most common applications of the term involve connecting normally non-current-carrying metal parts (conduit, boxes, etc.) to each other, resulting in a conductive path that will safely return accidental voltage to the source, usually opening the overcurrent device.

Ground wire This is the wire that runs from the service equipment to the grounding electrode (which is grounded because it is buried in the earth). The *NEC* calls it the "grounding electrode conductor." It is usually bare, but can be insulated and of any color. The ground wire is bonded within the service equipment enclosure, to the service raceway, to the service enclosure, and to the equipment grounding bus if any, and to the neutral (grounded) conductor on all grounded systems.

Grounded (white) wire In a circuit, this is the wire (usually white) that normally carries current and is connected to the *ground* at the service equipment. The *NEC* calls it the "identified conductor." The ground*ed* wire must never be fused or protected by a circuit breaker or interrupted by a switch. It will also be a *neutral*.

Grounding (green or bare) wire This is the *equipment grounding conductor*. It is *bonded* to components of the installation that normally do not carry current but do carry current in case of damage to or defect in the wiring system or the appliances connected to it.

The ground*ing* wire runs with the current-carrying wires. It must be green, green with one or more yellow stripes, or bare. It must never be used for any purpose except as the ground*ing* wire in a circuit. In the case of wiring with metal conduit, or cable with a metal armor, a grounding wire as such might not be installed separately because the conduit or the armor of the cable are eligible to serve as the ground*ing* conductor where it is bonded to the cabinet. Take care on this point, not all metallic cable or raceway armors qualify as equipment grounding conductors. For example, most common types of MC cable, as well as flexible and liquidtight flexible metal conduits over 6 feet in length (and even shorter lengths when used with circuits above certain ratings), are not suitable for use without a separate grounding conductor.

GROUNDING OVERVIEW

Proper grounding involves every part of the installation, including the incoming service wires that run from the power supplier's transformer to the building to be served. The service equipment cabinet—the heart of your electrical installation—is where the main grounding connections are made.

Wires from transformer to building   Figure 7-1 shows the power supplier's three wires—labeled *A*, *B*, and *N*—that run from the transformer to the building. The grounded wire, labeled *N*, is grounded both at the transformer and at the building's service equipment. This is a neutral wire (meaning it is non-current-carrying when the load is balanced on the hot wires). Wires *A* and *B* are "hot" wires. (Hot wires carry voltage above zero, or ground. One is usually black and the other red, or both black, but never white or green.) The voltage between *A* and *N*, or between *B* and *N*, is 120 volts; between *A* and *B* it is 240 volts.

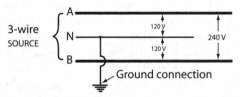

Fig. 7-1 The power supplier's three wires provide two different voltages. Use the lower voltage (120 V) for low-wattage loads, and the higher voltage (240 V) for high-wattage loads such as range and water heater. Note the symbol for a connection to ground.

The midpoint connection for the neutral wire within the transformer creates a neutral point, and all wires connected to this point are neutrals under new definitions as of the 2008 *NEC*. This remains true even on a two-wire circuit connected to this system, which is a reversal of prior usage.

Grounding of 120-volt and 240-volt loads   The grounded wire must run without interruption to all equipment operating at 120 volts, but not to anything operating only at 240 volts. Only hot wires run to 240-volt loads. A separate grounding wire runs to 240-volt loads (unless the conduit or the armor of armored cable serves as the grounding conductor). *Note:* Anything that is connected to a circuit and consumes power constitutes a "load" on the circuit. The load might be a motor, a toaster, a lamp—anything consuming power. Switches do not consume power and therefore are not loads. A receptacle is not a load, but anything plugged into the receptacle is a load.

Grounding at the service equipment   At the service equipment all the grounding wires are connected to an equipment grounding busbar, which is bonded to the enclosure, the grounding electrode conductor, and, for grounded systems within the scope of this book, connected to the neutral busbar (for all the white wires). This connection between grounded and grounding conductors is made through what the *NEC* defines as the *main bonding jumper*. It is the most important single connection in a grounded system. If it is not made, lethal voltages can remain on electrical equipment in the event of an insulation failure. It must be made at the service equipment and, with rare exceptions, this connection must never be made again at any downstream location in the system. This is the "single-point grounding" concept. It prevents ordinary load currents from flowing over electrical system components that were never designed to carry load current, such as switch enclosures, etc. For service equipment only, this concept means that equipment grounding and grounded conductors may land on the same busbar, a practice forbidden at all downstream locations.

Short circuits and ground faults   If two hot wires with a voltage between them touch each other at a point where both are bare, or if a hot wire touches a bare point in a grounded circuit wire, a short circuit occurs at the point of connection. This rarely happens in the actual wires of a properly installed system, but often happens in cords to lamps or appliances, especially if the cords are badly worn or abused.

A ground fault results when a bare point in a hot wire, such as where the wire is connected to a receptacle or switch, touches a grounded component such as conduit, the armor of armored cable, or a ground*ing* wire. For both short circuit and ground fault the effect is the same: a fuse will blow or a circuit breaker will trip.

HOW GROUNDING PROMOTES SAFETY

The diagrams in this section illustrate correct installations as well as faulty conditions in which shock can occur. The motors represented could be either freestanding or part of an appliance. The coiled portion of the line on the right side of each diagram represents the motor's winding. Your risk of shock and danger from a faulty installation will depend on the surface on which you are standing, your general physical condition, and the condition of your skin at the contact point. If you are on an absolutely dry surface you will note little shock. If you are on a damp surface (as in a basement) you will experience a severe shock. If you are standing in water you will undergo extreme shock or death. As a safety precaution, always stand on dry

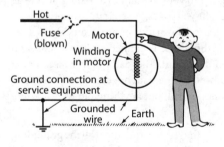

◀ **Fig. 7-2** A 115-V motor properly installed except for a grounding wire. With no grounding wire, the installation is safe *only as long as the motor remains in perfect condition.*

Fig. 7-3 A 115-V motor with a fuse wrongly placed in the grounded wire. It is a dangerous installation. ▶

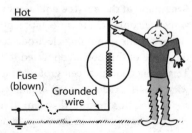

boards when you must work in a damp or wet location, and unless highly trained, always disconnect that portion of the wiring.

Grounding and fuse/breaker placement for 120-volt circuit   Figure 7–2 shows a 115-volt motor with the ground*ed* wire connected to the grounded neutral of the service equipment and a fuse in the hot wire. (Actually the fuse and the ground connection would be at the service equipment cabinet a considerable distance from the motor, not near the motor as shown, although there might be an additional fuse near the motor.) If the fuse blows, the motor stops. What happens if, while inspecting the motor, you accidentally touch one of the wires at the terminals of the motor? Nothing happens because the circuit is hot only up to the blown fuse. Between the fuse and the motor, the wire is now dead just as if the wire had been cut at the fuse location. The other wire to the motor is grounded, so it is harmless. You are protected. But if the fuse is *not* blown and you touch the hot wire, you will receive a 120-volt shock through your body to the earth, and through the earth back to the neutral wire at the service equipment. Note that, although the circuits in these motor examples are correctly classified as either 120-volt or 240-volt circuits, the motors are classified as 115-volt or 230-volt equipment by the *NEC*. Therefore, the description of circuit voltage differs slightly from the voltage rating applicable to the motor itself.

Fuse/breaker placement error   The same circuit is shown in Fig. 7–3, except the fuse is wrongly placed in the grounded wire instead of the hot wire. The motor will operate properly. If the fuse blows, the motor stops. But the circuit is still energized through the motor and up to the blown fuse. If you touch one of the wires at the motor, you complete the circuit through your body, through the earth, to the neutral wire in your service equipment; you are directly connected across 120 volts and as a minimum you will receive a shock and, at worst, will be killed.

Accidental internal ground—unprotected   Now see Fig. 7–4, which again

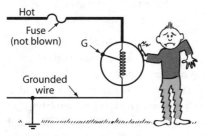

◄ **Fig. 7–4** The same motor as in Fig. 7–2, but the motor is defective. G represents an accidental grounding of the winding to the frame. The grounding wire has not been installed. This is a dangerous installation.

Fig. 7–5 The same defective motor as Fig. 7–4, but now a grounding wire has been installed from the frame of the motor to ground. Even though the winding is accidentally grounded to the frame, as represented by G, there is no shock hazard. ►

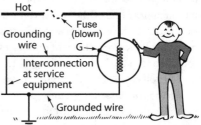

shows the same 115-volt motor as in Fig. 7–2. But suppose the motor is defective so that at the point marked *G* the winding inside the motor accidentally comes into electrical contact with the frame of the motor. As a result, the winding "grounds" to the frame. That does not prevent the motor from operating. But suppose you choose to inspect the motor, touching just its frame. What happens? Depending on whether the internal ground between winding and frame is at a point nearest the grounded wire or nearest the hot wire, you will receive a shock up to 120 volts as you complete the circuit through your body back to the ground*ed* wire. It is a potentially dangerous situation since shocks of much less than 120 volts can be fatal.

In fact, it is not uncommon for breakdowns in the internal insulation of a motor to result in an accidental electrical connection between the winding and the frame of the motor. The entire frame of the motor becomes hot. The same situation develops if there is an insulation failure in the supply to the motor in its terminal housing due to an abraded cord or poor splice.

Accidental internal ground—protected   Now see Fig. 7–5 which shows the same motor as in Fig. 7–4 with the same accidental ground between winding (or cord) and frame, but protected by a grounding wire that is connected to the frame of the motor and runs back to the ground connection at the service switch. When the internal ground occurs, current will flow over the grounding wire, and the amount of current that flows over is dependent upon the degree of insulation breakdown within the motor housing. If installed correctly, with no significant impedance in the path back to the system source, the fuse will blow immediately. *Although this is an equipment grounding conductor, its proper function does not depend on a connection to earth; it depends only on the connection to the grounded conductor at the service entrance.*

Although it is grounded (connected to earth) for other reasons, its usual protective function does not depend on this. The grounding electrode is not relevant to how this wire functions in this regard, and *the earth must never be used in place of a properly installed grounding wire.* Many ground rods have been installed at remote locations with no equipment grounding return path in the mistaken belief that "grounding" the remote equipment in this way provides shock protection.

Grounding and fuse/breaker placement for 240-volt circuit   Now refer to Fig. 7–6, which shows a 230-volt motor installed with each hot wire protected as required with a fuse or circuit breaker. Remember that in such 240-volt installations, the white ground*ed* wire does not run to the motor, but is nevertheless grounded at the service equipment. If you touch both hot wires, you will be completing the circuit from one hot wire to the other, and you will receive a 240-volt shock. But if you touch only one of the wires, you will be completing the circuit through your body, through the earth, back to the grounded neutral in your service equipment; and you will receive a shock of only 120 volts: the same as touching the ground*ed* wire and one of the hot wires of Fig. 7–1. The difference between shocks of 120 volts and 240 volts can be the difference between life and death.

Accidental internal ground—protected   Assume that the motor in Fig. 7–6 becomes defective, that the winding of the motor becomes accidentally grounded to

Hot

Hot

Ground at service
equipment

◀ **Fig. 7-6** A 230-V motor with each hot wire correctly protected by a fuse/breaker. The grounded wire does not run to 240-V loads. Because the grounding wire has not been installed, shock hazard is avoided *only if the motor remains in perfect condition.*

Fig. 7-7 This is the same installation as Fig. 7-6, but the motor is now defective. The winding is accidentally grounded to the frame, as represented by G. This is a dangerous installation. ▶

Hot

Hot

Ground at service
equipment

the frame, as shown in Fig. 7–7. This is the same as Fig. 7–4, except that the motor is operating at 240 volts instead of 120 volts. Touching the frame will produce a 120-volt shock. But if the frame has been properly grounded, as shown in Fig. 7–8, at least one, and likely both, of the fuses will blow. If one does not blow because of high impedance between the hot conductor and the point of the fault, touching the frame will not produce as dangerous a shock. This is because the frame is grounded with a voltage to ground, reflecting the product ($E=IR$) of the small current flowing on that leg due to high impedance, and the small resistance in the path between the motor frame and the circuit source.

GROUNDING CONSIDERATIONS FOR WIRING SYSTEMS

In any of the situations of Figs. 7–2 through 7–8, if there is an accidental contact between any two wires of the circuit, the contact constitutes a short circuit, and a fuse will blow or the circuit breaker will trip regardless of whether the short is between one of the hot wires and the grounded wire, or between the two hot wires. For a short circuit to occur, there must be bare places on two different wires touching each other, which does not happen very often in a carefully installed job.

Metal conduit or armored cable   The grounding conductors that make these motor installations safe are often, but not always, actual wires. The *NEC*, in 250.118, lists fourteen acceptable equipment grounding conductors, of which only the first is a separate wire (covered below). Other methods include three rigid metal raceways, three flexible metal raceways, three types of flexible cable with metallic armor, cable trays and cablebus framework, surface metal raceways, and other listed raceways, including metal wireways and similar enclosures, such as metal auxiliary gutters. In every instance, where the wiring method is bonded at both the source and the motor enclosure, the installation will safely clear any ground fault from continuing.

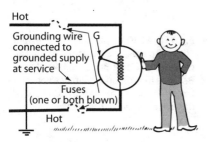

◀ **Fig. 7–8** The same defective motor as Fig. 7–7, but a grounding wire has been installed from the frame of the motor to ground. Even though the winding is accidentally grounded to the frame, as represented by G, the shock hazard is minimized.

Fig. 7–9 Wires in metallic conduit or metallic armor are used to install the motor. The conduit is grounded both at the service equipment cabinet and to the motor. It is a safe installation. ▶

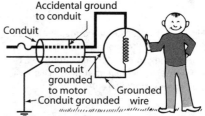

See Fig. 7–9. Remember the caution on page 66, however: not all metal raceways or cable assemblies qualify as equipment grounding conductors; and those that don't will require separate grounding conductors, just as in the case of nonmetallic wiring methods.

Separate grounding conductors   Many wiring methods use nonmetallic cable sheaths or raceways. In most such cases you install (or the cable includes) a separate grounding conductor that performs the same equipment grounding function as the steel raceway or cable armor. In addition, some designers insist on a separate equipment grounding conductor even within a steel raceway in order to add additional reliability. The *NEC* actually requires that arrangement for branch circuits serving patient care areas of hospitals. Although hospital wiring is beyond the scope of this book, that rule points to one of the many instances where you're likely to encounter separate grounding conductors.

Continuous grounding   As explained on the first page of Chapter 9, an outlet box or switch box is used to protect each electrical connection in an installation. When a grounded-neutral wiring system is used (which is 100 percent of the time for installations of the kind discussed in this book), and you use metal conduit or cable with a metal armor, you must ground not only the neutral wire but also the conduit or cable armor. The white wire is grounded only at the service equipment (at the point where it is connected to the neutral of the service equipment), but the conduit or armor must be securely connected to every box or cabinet. Lighting fixtures installed on metal outlet boxes are automatically grounded through the conduit or armor. (Nonmetallic boxes are also used; see Figs. 9–17 and 9–18.) Motors or appliances directly connected to conduit or armor are automatically grounded. If using nonmetallic-sheathed cable, the extra wire—the bare uninsulated grounding wire—must be carried from outlet to outlet, providing a continuous ground. The connections for this wire are explained on page 128-126. Regardless of the wiring method used, a continuous ground all the way back to the service is essential. Good workmanship is crucial for this part of the job, because a small defect in a grounding return path can make the difference between an overcurrent device opening promptly or not, leading to a fire or worse.

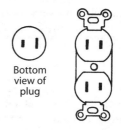

Fig. 7-10 Polarized receptacle and plug, with the wider slot connecting to the grounded conductor. A polarized plug cannot be inserted into a nonpolarized receptacle, so replace an old nonpolarized receptacle with the polarized type.

GROUNDING-TYPE RECEPTACLES INCREASE SAFETY

Many old receptacles still in use have only two parallel openings for the plug, as in Fig. 7-10. Plugging something into this receptacle duplicates the condition of Fig. 7-2. If you handle a defective appliance that is plugged into an old-style receptacle, you could receive a shock (See Fig. 7-4). This danger led to the development of the "grounding receptacle," which has three openings (See Fig. 7-11). Note that the grounding receptacle in Fig. 7-11 has the usual two parallel slots for two blades of a plug, plus a third round or U-shaped opening for a third prong on the corresponding plug.

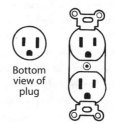

Fig. 7-11 Grounding receptacle and plug. Plugs with either 2 or 3 blades will fit.

In use, the third prong of the plug is connected to a third or grounding (green) wire in the cord, running to and connected to the frame of the motor or other appliance.

Making the ground connection   On the receptacle, the round or U-shaped opening leads to a green terminal screw that in turn is connected to the metal mounting yoke of the receptacle. The details of how to connect the green terminal to the equipment grounding conductor, or to the grounded metal box, are discussed in Chapter 11 on pages 128-128. In this way, the frame of the motor or appliance is effectively grounded, leading to extra safety as shown and discussed in connection with Fig. 7-5.

Where grounding receptacles are required   If an appliance is connected by cord and plug, *NEC* 250.114(3) requires a three-wire cord and three-prong plug on every refrigerator, freezer, air conditioner, clothes washer, clothes dryer, dishwasher, sump pump, on aquarium equipment, and on every hand-held, motor-driven tool such as a drill, saw, sander, hedge trimmer, and similar items. The three-wire cord with three-prong plug is not required on ordinary household appliances such as toasters, irons, radios, TVs, razors, lamps, and similar items.

Only one kind of receptacle needs to be installed, because grounding receptacles are designed for use with both two-prong and three-prong plugs. The *NEC* requires that in all new construction only grounding receptacles are to be installed. (See page 223 for replacing two-wire receptacles in existing installations.)

GROUND-FAULT CIRCUIT INTERRUPTER (GFCI)

The GFCI is a supplementary protection that senses leakage currents too small to operate ordinary branch circuit fuses or circuit breakers. The use of a grounding wire in a three-wire cord with a three-prong plug and a grounding receptacle reduces the danger of a shock in some circumstances, such as when using a portable tool, but it does not eliminate the danger completely. Cords can be defective or wrongly connected. Millions of tools with two-wire cords are still in use. Some people foolishly cut off the grounding prong on a three-prong plug because they have only two-wire receptacles (See Figs. 7–10 and 7–11).

Under normal conditions the current in the hot wire and the current in the grounded wire are absolutely identical. But if the wiring or a tool or appliance is defective and allows some current to leak to ground, perhaps giving someone a shock in the process, then a ground-fault circuit interrupter will sense the difference in current in the two wires. If the fault current exceeds the trip level of the GFCI, which is between 4 and 6 milliamperes, the GFCI will disconnect the circuit in as little as one-fortieth of a second.

The GFCI should be considered additional insurance against dangerous shocks. It is not to be considered a substitute for grounding. The GFCI will not prevent a person who is part of a ground-fault circuit from receiving a shock, but it will open the circuit so quickly that the shock will be below levels that inhibit breathing or heart action or the ability to let go of the circuit.

Where GFCIs are required   *NEC* 210.8 and other specialized rules require GFCI protection for 15-amp and 20-amp receptacles in these locations in dwellings: in bathrooms, at all kitchen counters, within 6 feet of sinks in all other locations, outdoors, in grade-level nonhabitable rooms of detached accessory buildings and in garages, in unfinished basements and crawl spaces, and in boathouses. The protection must be provided for boat hoist circuits (even hard-wired) in any occupancy, and for receptacles used temporarily during construction. GFCI protection is also generally required for commercial and institutional bathroom, outdoor, and kitchen receptacles wherever located, and for receptacles within 6 ft of sinks, in repair garages, in indoor wet locations, and in locker rooms with showers.

Three types of GFCI   Ground-fault circuit interrupters are available as separately enclosed types, or in combination with either a breaker or a receptacle.

▓ The *separately enclosed type* is available for 120-volt, two-wire and 120/240-volt, three-wire circuits up to 60 amps. It is most often used in swimming pool wiring, installed at any convenient point in the circuit. Related to this is the "master trip" GFCI device, which looks like a GFCI receptacle without any slots; it is used for downstream GFCI protection in cases where protection must be provided, but an outlet is not desired or permitted at its location.

■ The *breaker type* combines a 15-amp, 20-amp, and up to 60-amp circuit breaker and a GFCI in the same plastic case. It is installed in place of an ordinary breaker in your panelboard, and is available in 120-volt, two-wire or 120/240-volt, three-wire types (which will protect a 120/240-volt, three-wire circuit or a 240-volt, two-wire circuit). It provides protection against ground faults and overloads for all the outlets on the circuit. You can at any time replace an ordinary breaker in your panelboard with one of these combination breakers. Each GFCI circuit breaker has a white pigtail that you must connect to the grounded (neutral) busbar of your panelboard. You must connect the white (grounded) wire for the circuit to a terminal provided for it on the breaker. Do not use a single-pole GFCI circuit breaker on a multiwire circuit; it will nuisance trip immediately. Use two-pole GFCI circuit breakers on such circuits.

■ The *receptacle type* combines a receptacle and a GFCI in the same housing. It provides only ground-fault protection to the equipment plugged into that receptacle or, if it is the "feed through" type, to equipment plugged in to other ordinary receptacles installed "downstream" on the same circuit. This type is a convenient choice when replacing existing receptacles where GFCI protection is desired or required. Be very careful to observe "line" and "load" markings on these receptacles. If you wire them backwards, anything plugged in will still work, and the test button will cause the reset button to trip as usual. However, only downstream loads will be protected; anything plugged into the miswired device will have no GFCI protection at all.

GFCI testing and potential problems   Regardless of the type or brand of GFCI you install, it is essential that you carefully follow the installation and periodic testing instructions that come with it. Every GFCI has a test button for easy verification of its functional operation.

The GFCI is designed to trip if the cords or tools plugged into the protected receptacle outlet are in poor repair and provide a path for current to leak to ground. Even where wiring, tools, and appliances are in perfect condition and there is no ground fault, be on the lookout for these installation problems that will cause tripping of a GFCI:

■ A two-wire GFCI receptacle (other than an end-of-run application) is connected in a three-wire circuit. Two-pole GFCI circuit breakers will protect these circuits.

■ The white circuit conductor is grounded on the load side of the GFCI.

■ The protected portion of the circuit is excessively long (250 feet maximum is the rule of thumb—longer circuits may develop a capacitive leakage to ground).

Remember that the GFCI will not prevent shock, but it will render shocks relatively harmless. Also, it will not protect a person against contact with both conductors of the circuit at the same time unless there is also a current path to ground.

SURGE PROTECTIVE DEVICES

With the increasing home use of personal computers and other sensitive electronic equipment on general-use branch circuits, there is a need for suppression of voltage surges. These surges are typically of very short duration. Among the possible causes are the switching off or on of fluorescent lights or motors such as air conditioners, or the switching of major loads by the utility, or distant electrical storms.

The term "surge protective device" is, as of the 2008 *NEC*, the term of art for devices that shunt voltage spikes to ground. At one time the term was "lightning arrester," but, effective with the 1981 *NEC*, the terminology changed to "surge arrester" in recognition that voltage surges had many origins in addition to lightning. The next step occurred in the 2002 *NEC*, when a new article recognized "transient voltage surge suppressors" in addition to surge arresters. These devices were normally located on the load side of the service equipment, but could be on the line side if special arrangements were made. This has changed, and the terminology "surge arrester" is now reserved for medium-voltage (over 600V) applications exclusively. The term "surge protective device" covers the entire installation spectrum for 600V and lower applications.

The *NEC* recognizes three types of surge protective devices. Type 1 devices correspond to the old surge arresters and are suitable for installation on the line side of the service equipment. Type 2 devices can be installed at any point on the load side of a service disconnect; the photo in Fig. 17-8 shows an example in a service panel. They are also permitted on the load side of the first overcurrent device in a building or other structure supplied by a feeder. When special arrangements are made for overcurrent protection, these devices can also go on the line side of the service. Type 3 devices, the least robust, are permitted only on the load side of branch circuit protective devices and with the further restriction that they are at least 30 ft, measured along the conductors, from the service or local feeder disconnect for the building.

Several types of surge suppressors are available including:

- a plug-in unit resembling a cord adapter
- a group of receptacles on a strip with a supply cord
- permanently installed receptacles that include surge suppression in their design
- a unit shaped like a circuit breaker that provides only surge suppression— it is plugged into a two-pole circuit breaker space
- circuit breakers which, in addition to the usual overcurrent protection features, incorporate a surge suppressor to protect the entire branch circuit

To afford complete protection, a combination of several devices must be used, steadily reducing both the transient voltages and the let-through energy to which your equipment may be exposed. How much you spend will follow from the complexity of the application and component quality. Be sure the unit you purchase is listed.

PART 3 INSTALLING SERVICE EQUIPMENT AND WIRING

Chapter 8
THE SERVICE ENTRANCE

ELECTRIC POWER FLOWS FROM THE supplier's transformer through a set of three wires to your service entrance where the wires enter the building. This chapter discusses the design and installation of a service entrance for a house. Installations for farm buildings are discussed in Chapter 17.

PLANNING YOUR SERVICE ENTRANCE

These are the components of the service entrance (shown in use in Fig. 8–6).
Outside the building—

■ Service head, also called the entrance cap or weather head

■ Insulators anchored to the building below the service head—this is where the power supplier's wires end

■ Service entrance cable or wires in conduit leading from the insulators to the service head and down to the meter socket

■ Meter socket

■ Service entrance cable or wires extend from the meter socket into the building

Inside the building—

■ Service entrance cabinet that contains the disconnecting means and overcurrent protection—either circuit breakers or fuses in a panelboard

■ Ground connection—a ground wire from the incoming service to the busbar in the cabinet and back outside the building to the grounding electrode system

Three-wire service entrance The standard installation is the three-wire entrance. When three wires are installed, both 120 volts and 240 volts are available as shown in Fig. 7–1. The voltage between N and A, or between N and B, is 120 volts; between A and B it is 240 volts. The wire N is white (sometimes bare) beginning at the insulators. It is grounded, and up to the service equipment inside the building it is a *neutral* wire. The two hot wires can be any color except white or green, but are usually black and red, or both black.

Service equipment rating   Service circuit breakers are rated at any of the standard sizes for circuit breakers, but the enclosure must be rated as suitable for use as service equipment. Service fusible switches are rated at 30, 60, 100, and 200 amps, although any standard-size fuse may be used within such a switch, provided the switch is rated as suitable for use as service equipment. Equipment larger than 200 amps is available for larger installations. Some power suppliers will furnish a self-contained meter that will handle a 400-amp service, but beyond that the meter must be served by current transformers in a separate enclosure ahead of the service disconnect.

The service equipment of a single-family dwelling is required to have a rating of 100 amps or higher. In wiring buildings other than houses, a 30-amp service may be installed if there are not more than two 2-wire circuits. In all other cases, the equipment must be adequate for the load involved, but never less than 60 amps. See 230.79 in the *National Electrical Code (NEC)*. Note that feeders to individual dwelling units in an apartment house have essentially no arbitrary minimum size; they are simply sized to meet the load.

To determine the maximum wattage, or volt-amperes (VA), available with any size switch or breaker, multiply the amperage rating by the voltage (240). For example, a 60-amp switch or breaker will make available 14,400 VA ($60 \times 240 = 14,400$), and a 100-amp size will make available 24,000 VA ($100 \times 240 = 24,000$).

What size service?   To determine the size of service you need, begin by computing the total probable load in volt-amperes (net computed VA).

A 100-amp service is the smallest permitted by the *NEC*. The 100-amp service provides a maximum capacity of 24,000 VA. If the total of the "net computed VA" that you just calculated comes to more than 24,000 VA, does it follow that a service larger than 100 amps is necessarily required? No, because all the lights and appliances listed in the tabulation will not be in use at the same time. But remember that the assortment of appliances you include in your first tabulation represents your present ideas. As time goes on, you will add more appliances. A 150-amp or 200-amp service will meet future needs more adequately. In some areas, local codes are already requiring services larger than 100 amps. The larger services cost relatively little more than the 100-amp size.

Service entrance wire size   Since 100-amp service is the minimum for homes, select a wire with ampacity of 100 or more amperes from Table 4–1 on page 28. For installations other than homes, there is no arbitrary *NEC* minimum, but *NEC* 230.42 does require, for any service, that the conductors be sized to carry the load as calculated in Article 220. Examples of dwelling service calculations are found in *NEC* Annex D, Examples D1a, D1b, D2a, D2b, and D2c. Remember the note on page 28 about using Column B of the table for this work. However, *NEC* 310.15(B)(7) has a special table for dwelling services (or main feeders in multifamily dwellings) that has the effect of decreasing by, typically, one wire size from the normal sizing practices. For example, 4 AWG copper is permitted for a 100-amp dwelling service.

CALCULATING TOTAL PROBABLE LOAD IN VOLT-AMPERES

Area of house* _____ square feet × 3 VA . . . . _____ VA

Two appliance circuits . . . . 3,000 VA†

One laundry circuit . . . . 1,500 VA

Range, if used . . . . _____ watts‡

Water heater, if used . . . . _____ VA

Other permanently connected appliances . . . . _____ VA

Total . . . . _____ VA

*As calculated in Chapter 5 on page 48.

†Each of these circuits will provide 2,400 VA (20 × 120 = 2,400), but for the purpose of this calculation use 1,500 VA per circuit (3,000 VA altogether) because it is not likely that both circuits will be loaded to capacity at the same time. If more such circuits will be provided in any particular installation, multiply the number of such circuits by 1,500 VA.

‡Per the *NEC*, allow 8,000 watts if the rating of your range (or countertop unit plus separate oven) is 12,000 watts or less. If rated over 12,000 watts, add 400 watts for each additional kilowatt or fraction of a kilowatt. For ranges, watts and VA are considered equivalent. For examples and further details, see Annex D in the 2014 *NEC*.

DISCONNECTING MEANS

At the service location you must provide equipment, a main switch or circuit breaker, usually in a panelboard, which allows you to disconnect all the wiring from the power source. You must also provide either circuit breakers or fuses to protect the installation as a whole as well as each branch circuit individually. The service disconnect may be either inside or outside. In conjunction with the discussion that follows, be sure to review Chapter 5, "Circuit Protection and Planning."

Advantages of circuit breakers   Most installations use circuit breakers because of their many advantages over fuses. If a breaker trips, you can restore service by merely flipping its handle (after correcting the problem that created the overload that tripped the breaker—see pages 218–221). You can also disconnect an individual circuit by using the breaker as you would a switch. The breakers will carry nondangerous temporary overloads that would blow ordinary fuses.

Selecting breaker equipment   Most service entrance equipment as purchased will contain one large (100-amp or larger) two-pole main breaker that will protect the entire installation and will let you disconnect the entire installation from the power source by merely turning this one breaker to the OFF position.

But there must be more breakers to protect individual circuits. You will need a single-pole breaker for each 120-volt circuit and a two-pole breaker for each 240-volt circuit. These branch-circuit breakers are not part of the equipment as purchased, but the cabinet contains an arrangement of busbars into which you can plug as many individual single-pole and two-pole breakers as there is room for. A two-pole breaker occupies the same space as two single-pole breakers. Each panelboard

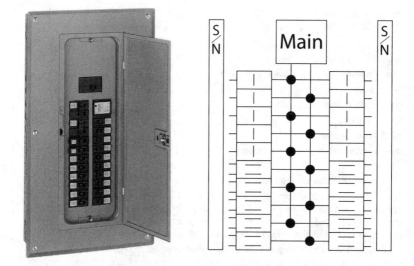

Fig. 8–1 On the left, a circuit breaker cabinet with main breaker *(Square D Company)*. On the right, diagram showing space for 20 circuits. *(Square D Company)* Arrangements for mounting plug-in circuit breakers vary by manufacturer; a label in the panelboard lists catalog numbers of breakers permitted to be installed.

accommodates only as many breakers as its listing permits. The overall rating of the panelboard must equal or exceed the load served as calculated by *NEC* Article 220. Figure 8–1 shows a circuit breaker cabinet with its internal wiring diagram. The diagram shows space for 20 circuits. Select equipment that provides for the number of branch circuits you need plus a few additional spaces for future expansion. For help in calculating your circuit needs see the "What size service?" discussion earlier in this chapter and Table 5–3.

Split-bus breaker equipment Somewhat less popular are "split-bus" service entrance panelboards. The *NEC* permits two main disconnects in one service panelboard, provided the number of load-side overcurrent device positions does not exceed 42. To disconnect everything from the power source, it is necessary to turn *both* of them to the OFF position. In addition to the main breakers, the cabinet may contain as many breakers as you wish (up to the capacity of the panelboard), single-pole or two-pole, each protecting its own circuit, but all of these smaller breakers must be protected by one of the main breakers. See wiring diagram of Fig. 8–2. An alternate arrangement would be for one of the main breakers to protect all the branch circuits in the service panelboard, and the other main breaker to supply a feeder to a panelboard located elsewhere.

Outdoor cabinet In a few areas, it is common practice to use an outdoor cabinet containing the meter socket and the service disconnects (one to as many as six). Generally, major loads are split out of these panels, and one of the breakers supplies a remote panelboard in the basement or elsewhere that carries all the 15- and 20-amp. breakers for lighting and small appliance loads. This arrangement is

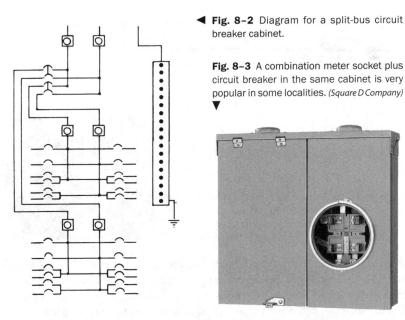

◀ **Fig. 8-2** Diagram for a split-bus circuit breaker cabinet.

Fig. 8-3 A combination meter socket plus circuit breaker in the same cabinet is very popular in some localities. *(Square D Company)*
▼

only permitted as part of service equipment. Figure 8–3 shows a meter socket and single main disconnect in one cabinet. In mild climates, all the branch circuits could also originate in this enclosure.

Fused equipment   This equipment is common in older homes, but almost impossible to find at supply houses, having been replaced by circuit breakers for new installations. The type widely used in residential work has two main fuses mounted on a pullout block (Fig. 8–4). With the pullout in your hand, insert the fuses into their clips. One side of each fuse clip has long prongs; the entire pullout has four such prongs. The equipment in the cabinet has four narrow open slots, but no exposed live parts; the live parts are behind the insulation. Plug the pullout with its fuses into these slots; the four prongs on the pullout make contact with the live parts, completing the circuit. Plugging the pullout into its holder is the same as closing a switch with hinged blades; removing it is the same as opening a switch. You can insert the pullout upside-down to leave the power turned off.

Such fused service equipment usually contains one main pullout with large fuses protecting the entire load, plus additional pullouts to protect 240-volt circuits with large loads such as range, clothes dryer, and water heater. There are also as many fuseholders for *plug* fuses as are needed to protect individual branch circuits: one for each 120-volt circuit, two for each 240-volt circuit. A service panelboard of this type is shown in Fig. 8–5. The wiring diagram is the same as in Fig. 8–1 except that fuses are used in place of breakers.

Fuseholders   For fuses of the cartridge type, the "60 amp" holder accepts fuses rated from 35 amps to 60 amps. The "100 amp" holder accepts fuses rated from 70

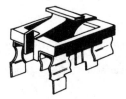

◀ **Fig. 8-4** In fused service equipment, cartridge fuses are usually installed on pull-out blocks. Replace fuses while the block is in your hand.

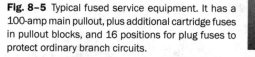

Fig. 8-5 Typical fused service equipment. It has a 100-amp main pullout, plus additional cartridge fuses in pullout blocks, and 16 positions for plug fuses to protect ordinary branch circuits.

amps to 100 amps. The "200 amp" holder accepts fuses rated from 110 amps to 200 amps. Holders for plug fuses accept fuses up to 30 amps (the top rating for plug fuses). In new work, Type S adapters must be used; these allow installation of only the proper fuse to match the ampacity of the wires. Fusible equipment for individual loads is still available and commonly used, particularly for commercial and industrial applications. Fuses can be designed with certain electrical characteristics, including speed of operation and low-current size ranges, which are unavailable in circuit breakers.

Split-bus fused equipment As in the case of breaker equipment, the fused equipment *may* contain two *main* pullouts, connected internally so the service wires run directly to each. To disconnect the load completely you must pull out both the main pullouts. The branch-circuit fuses must be protected by one of the main fused pullouts. The wiring diagram of connections within the equipment is that of Fig. 8-2 except that fuses replace the breakers.

Old-style fused service equipment In very old houses you will find service equipment that consists of a main switch with two hinged blades and an external handle used to turn the entire load in the building on and off. The switch cabinet also contains two *main* fuses. Usually it also contains as many smaller fuses as needed to protect all the individual branch circuits in the house. Sometimes these fuses are in a separate cabinet. If your service is underground, refer to pages 194–196 and 197.

INSTALLING THE SERVICE ENTRANCE

So far we have looked at the major parts that make up the service entrance, and how to select the right kind and size of equipment. The rest of this chapter concerns the installation of these parts. Figure 8–6 shows the details of a typical service entrance.

You will have to install insulators on the outside of the building. The power supplier will run its wires up to that point and connect them there to the service-entrance wires that you provide. You can use special cable for that purpose or run wires through conduit. Your wires will go up to the service head, down to the meter socket, and then into the house to the service equipment.

After installing the service equipment, you must properly ground the installation. Then you will be ready to wire the branch circuits.

Service equipment location The *NEC* requires this equipment to be located as close as practical to the point where the wires enter the house. Service conductors are by definition without overcurrent protection, so a service fault must burn clear.

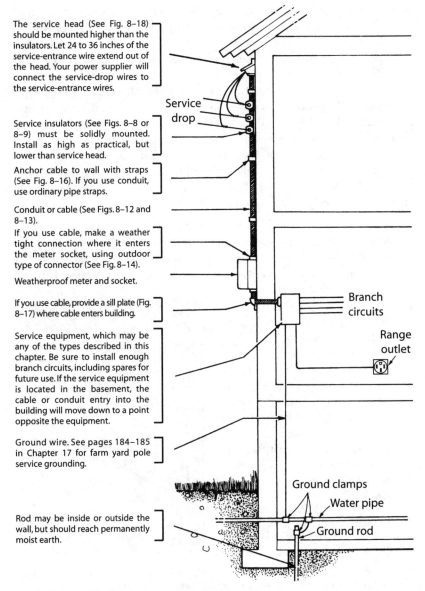

The service head (See Fig. 8–18) should be mounted higher than the insulators. Let 24 to 36 inches of the service-entrance wire extend out of the head. Your power supplier will connect the service-drop wires to the service-entrance wires.

Service insulators (See Figs. 8–8 or 8–9) must be solidly mounted. Install as high as practical, but lower than service head.

Anchor cable to wall with straps (See Fig. 8–16). If you use conduit, use ordinary pipe straps.

Conduit or cable (See Figs. 8–12 and 8–13).

If you use cable, make a weather tight connection where it enters the meter socket, using outdoor type of connector (See Fig. 8–14).

Weatherproof meter and socket.

If you use cable, provide a sill plate (Fig. 8–17) where cable enters building.

Service equipment, which may be any of the types described in this chapter. Be sure to install enough branch circuits, including spares for future use. If the service equipment is located in the basement, the cable or conduit entry into the building will move down to a point opposite the equipment.

Ground wire. See pages 184–185 in Chapter 17 for farm yard pole service grounding.

Rod may be inside or outside the wall, but should reach permanently moist earth.

Service drop

Branch circuits

Range outlet

Ground clamps

Water pipe

Ground rod

Fig. 8-6 A cross-sectional view of a typical service entrance.

The less the dwelling interior is exposed to unprotected wiring the better, which is the reason for the *NEC* rule. In other words, the service wires must not run 10 or 20 feet inside the house before reaching this equipment. Decide where to put your equipment cabinet before deciding where the wires are to enter the house. However, since the power supplier often specifies the meter location, as a practical matter you probably will have little to say about the service location. Remember also that the service wires may be run outside the building for any distance to the eventual point of entrance; this allowance can be used to accommodate an inconvenient meter location.

Service equipment must be readily accessible, and it must be in a space that is clear of any obstructions not a part of the electrical installation, both down to the floor and up as far as the structural ceiling. Special *NEC* rules and allowances apply if the ceiling is more than 6 feet above the service (or similar) equipment, however they are beyond the scope of this book [review *NEC* 110.26(F) if this applies to your situation]. In addition, there must be clear workspace around the equipment, at least 30 inches wide, 3 feet deep, and 6½ feet high, or higher if the equipment extends higher. Note that switch and circuit breaker operating handles, in general, cannot be higher than 6 ft 7 in. above the floor, allowing most individuals to operate them promptly in an emergency.

Meter and socket location   The outdoor type of meter is standard today. The weatherproof meter provided by the power supplier is plugged into a weatherproof socket (See Fig. 8-7). The socket is usually, but not always, provided and installed by the contractor, usually about 5 feet above the ground. The power supplier, however, will usually specify the meter location, and will have a list of meter sockets it will accept. In the case of indoor equipment, the meter socket is installed on a substantial board near the service equipment. For a two-family dwelling, two meter sockets are installed in one sheet-steel enclosure. Below the meters each occupancy will have its own service disconnect. In this way each tenant has access to their disconnecting means, and the disconnecting means for the building are grouped in one location, as required by the *NEC*.

Insulators   The incoming service wires are anchored on service insulators installed on the building. They should be installed a foot or so below the highest practical point so the wires to the inside

Fig. 8-7 Typical outdoor weatherproof kilowatthour meter and its socket.

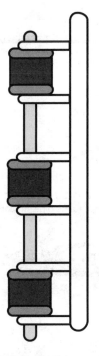

Fig. 8-8 Insulator rack for supporting three wires. Be sure to anchor rigidly to building.

Fig. 8-9 Screw-point insulator.

of the house, where they emerge from the service head on the conduit or cable, will slope downward toward the insulators to prevent water from following the service wires into the service head (See Fig. 8–6). Figure 8–8 shows a rack for individual service conductors, once common but rarely used today. Individual screw-point insulators (Fig. 8–9) are usually used, with the service drop containing all the service wires twisted together on a bare messenger wire, designed to accept strain, and serving as the grounded conductor in the drop.

Service wire clearance   *NEC* 230.24(B) requires that cabled service wires not over 150 volts to ground (or drip loops as illustrated in Fig. 8–19) be kept 10 feet above finished grade at the point of attachment to the building; 12 feet above residential property and driveways and commercial property not subject to truck traffic; 18 feet over parking areas and driveways other than residential, and farm properties that are subject to truck traffic. No wires may come closer than 3 feet to an openable window (there is no restriction to wires above the top of the window), or 3 feet horizontally and 10 feet vertically from any door, porch, fire escape, or similar location from which the wires might be touched. If they pass over all or part of a roof that has a rise of 4 inches or more per foot, they must be kept at least 36 inches from the nearest point. If the roof is flatter (less than 4 inches of rise per foot), they must be kept at least 8 feet from the nearest part of the roof.

Masts   In wiring a rambler or ranch-style house it is difficult to maintain required clearances if the insulators are mounted directly on the side of the house, so a mast is used. Several types are on the market, and Fig. 8–10 shows a typical construction. The service conduit extends upward through the roof overhang and becomes the support for insulators of the type shown in Fig. 8–11. Unless your power supplier has other requirements, the service conduit should not be smaller than 2-inch rigid metal conduit. Smaller conduit could be used if braced or guyed for extra strain support, or either smaller conduit or service entrance cable could be attached to a

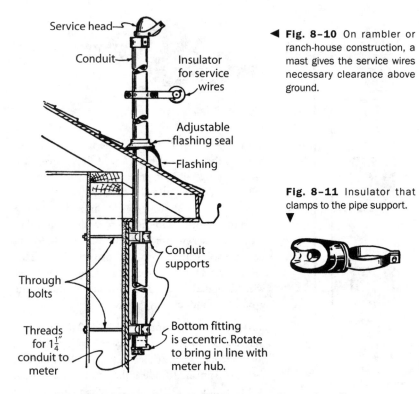

Service head

Conduit

Insulator for service wires

Adjustable flashing seal

Flashing

Through bolts

Conduit supports

Threads for $1\frac{1}{4}''$ conduit to meter

Bottom fitting is eccentric. Rotate to bring in line with meter hub.

◀ **Fig. 8–10** On rambler or ranch-house construction, a mast gives the service wires necessary clearance above ground.

Fig. 8–11 Insulator that clamps to the pipe support.
▼

4 × 4-inch timber. If the roof overhang does not exceed 48 inches, the service wires may be as low as 18 inches above the overhang.

Service-entrance wires   Service-entrance wires begin where the overhead (or underground) service wires end, and they stop at the service equipment where the individual circuit wiring begins. You may use service wire cable manufactured for the purpose, or separate wires in conduit.

Service-entrance cable   Typically, service-entrance cable (Fig. 8–12) is used to bring wires into the building. One of the wires is not insulated and consists of a number of fine wires wrapped around the insulated wires. In use, the small bare

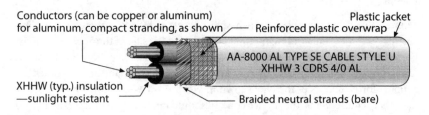

Conductors (can be copper or aluminum) for aluminum, compact stranding, as shown

Plastic jacket

Reinforced plastic overwrap

AA-8000 AL TYPE SE CABLE STYLE U
XHHW 3 CDRS 4/0 AL

XHHW (typ.) insulation —sunlight resistant

Braided neutral strands (bare)

Fig. 8–12 In service entrance cable the neutral wire is not insulated, but is bare and wrapped spirally around the insulated wires. The drawing shows 3-wire cable.

Strands of bare
wire, twisted

Fig. 8-13 When using the cable, the separate
strands of the neutral bare wire are gathered
into a bunch making one large wire.

wires are twisted together to make one larger wire (Fig. 8–13). The bare wire may
be used only for the grounded neutral. Over all is a fabric braid or outer protective
layer. This usually has a gray finish which may be painted to match the building.

Use cable with at least 2 AWG aluminum wires (or 4 AWG copper, but that
is much less common) with a 100-amp main breaker or switch in the case of a
single dwelling. Use cable with 4/0 AWG aluminum (or 2/0 AWG copper) wires
for a 200-amp breaker or switch at a similar building. In the case of commercial
or multifamily-dwelling applications, you will need to go up one wire size on all
100-amp applications, and on any 200-amp application where the calculated load
is higher than 180 amps. Services above 200 amps are beyond the scope of this
book. As explained on page 78, in houses 30-amp services and 60-amp services
are not permitted but may be used in other buildings such as a freestanding studio
or workshop on residential property. Use 8 AWG wire with a 30-amp breaker or
switch, or 6 AWG with a 60-amp breaker or switch.

If the installation served by the service equipment has no significant loads operating at 240 volts, all wires in the cable, including the neutral, must be the same size.
But if the installation has an electric range or water heater or other 240-volt loads
consuming about one-third of the total watts, then you may use cable with the neutral
one size smaller than the insulated wires—for example, 2 AWG insulated wires with
a 4 AWG bare neutral. Technically, the required calculation for the neutral must be
customized to the actual load profile and certain grounding return path issues, but
the rule of thumb presented here should be adequate for users of this book.

Wherever the cable enters a service equipment cabinet or a meter socket, it must
be securely anchored with a connector. Outdoors, a weatherproof connector must be
used; a common type is shown in Fig. 8–14. The connector consists of a body and a
heavy block of rubber, and a clamping nut or cover that compresses the rubber against

Fig. 8-14 Watertight
connectors for outdoor
use.

the cable, making a watertight joint. Don't assume that a connector is watertight just because it has the neoprene gland construction. Until 1998, test labs assumed that, due to variations in cable diameters, the connectors could not be assumed to be watertight. Standard trade practice has been to supplement the gland with sealing putty. Now, however, some of these connectors can pass the tests without supplementary waterproofing. They are identified by a separate RAINTIGHT marking on the carton. A less expensive connector of the type shown in Fig. 8–15 may be used indoors, or even outdoors if below the level of any internal uninsulated live parts (such as the bottom wall of a meter socket).

Fig. 8–15 An ordinary connector for indoor use.

Cable is anchored to the building with straps like the one in Fig. 8–16. At the point where the cable enters the building, you must take steps to prevent rain from following the cable into the building. You can do this by arranging the geometry of your lower cable run to shed all water prior to arriving at the service equipment, but the simplest and most workmanlike way to seal this point is to use a sill plate such as shown in Fig. 8–17. Soft waterproofing compound that comes with the plate is used to seal any opening that may exist. When you install service entrance cable, cut a length long enough to reach from the meter socket to a point at least a foot above the topmost insulator, plus another two or three feet. On this last additional length, remove the outer braid over the spirally wrapped neutral wire. Unwind these bare wires from around the cable and twist them into a single wire; see Fig. 8–13. Then install a service head of the general type shown in Fig. 8–18, letting the individual wires project through separate holes in the insulating block in the service head. The service head is designed to prevent water from entering the top of the cable. Anchor the service head on the building about 12 to 18 inches above the topmost insulator so that, after the connection has been made to the power supplier's wires, rain will tend to flow away from the service head rather than into it.

◀ **Fig. 8–16** Entrance cable must be anchored to the building every 30 inches. Use straps of the type shown.

Fig. 8–17 Use a sill plate packed with sealing compound where cable enters building. It keeps water out. ▶

◀ **Fig. 8–18** This service head is for cable. It prevents water from entering the cut end of the cable. A service head for conduit is supported by the conduit.

If it is impossible to locate the service head above the insulators, be sure to provide drip loops (See Fig 8–19) to keep rain out of the cable. Your power supplier may connect the ends of the wires in the cable to the service-drop wires, although local service policies vary. In some jurisdictions, particularly on upgrades of smaller residential services, the electrician on site must do the cut and reconnection. Because the service drop will be live, this is potentially dangerous work for an unqualified person to perform.

Support the cable within 12 inches of all terminations, and additionally every 30 inches using straps like those shown in Fig. 8–16. Run the bottom end of the cable into the meter socket, again twisting the neutral strands into a single wire. Use the same procedure to go from the socket to the service equipment. Make the connections as shown in Fig. 8–20, landing the twisted neutrals on the center contacts.

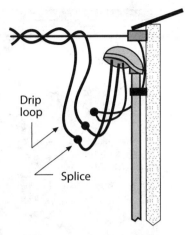

Fig. 8–19 If the service head cannot be located higher than the insulators, provide drip loops. Splice at bottom of loop, and insulate.

Separate wires in raceway (conduit)   Instead of service entrance cable, separate Type TW, THW, RHW, THWN, or XHHW wires inside raceway may be used. The neutral wire is always white (or reidentified as such using paint or tape at terminations); in some localities bare uninsulated wire is used. The other two wires are black, or black and red. To determine the size raceway required, see Tables 1, 4, and 5 in Chapter 9 of the *NEC*.

Common raceways include rigid or intermediate metal conduit, rigid nonmetallic conduit, and electrical metallic tubing, all described in Chapter 12. Cut pieces as long as required and prepare them in the manner described in Chapter 12 depending on the type used. At the top of the raceway install a service head of the type shown on the right in Fig. 8–21. At the point where the raceway is to enter the house, use an entrance ell such as shown on the left in Fig. 8–21. This has a removable cover which makes it easy to help the wires around the sharp bend while pulling them into the raceway. Be sure to get one large enough, as covered in the next topic.

After the raceway itself is completely installed, pull the wires into it. For the relatively short lengths involved, the wires can usually be pushed in at the top and down to the meter socket; from there, other lengths are pushed through to the inside of the house. If the raceway is long or has bends in it, use fish tape to pull the wires through. See Fig. 12–11 and related text for how to use fish tape. A length of galvanized clothesline or similar wire will serve the purpose for short lengths.

Conduit bodies   At the point where the raceway enters the building (which would correspond to the location of the sill plate), it is customary to use an entrance conduit body of the type shown in Fig. 8–21. With the cover removed, it is a simple matter to pull wires around the right-angle corner. Be careful to observe the sizing

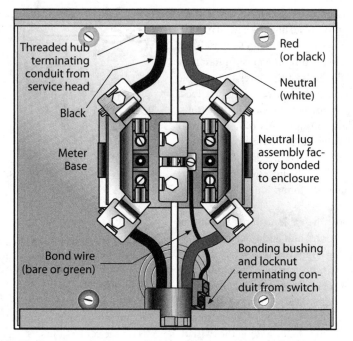

Threaded hub
terminating
conduit from
service head

Black

Meter
Base

Bond wire
(bare or green)

Red
(or black)

Neutral
(white)

Neutral lug
assembly fac-
tory bonded
to enclosure

Bonding bushing
and locknut
terminating con-
duit from switch

Fig. 8-20 How wires are connected to the meter base or socket. If cable is used, the bare uninsulated wire is always connected to the neutral center contact of the socket, and the bonding bushing and wire are not used.

rules in *NEC* 314.28(A)(2) and (3) in choosing this fitting. The distance between the center of the conduit stops (roughly equal to the long dimension of the cover) must not be less than 6 times the trade diameter of the entering raceways, or a full 12 inches in the case of trade size 2 raceway. The distance between the conduit stop in the short dimension and the inside of the cover must at least equal the minimum bending radius for the size of conductors used, based on *NEC* 312.6(A). For 3/0 copper conductors, rated 200 amps, that dimension is 4 inches. Conduit bodies (which generally look like the letter "L"; those with the opening in the back are called "LB") that meet these spacings are generally referred to as "mogul" fittings; conventional conduit bodies won't meet *NEC* requirements.

The only other option is to see if a conduit body of smaller dimension has

Fig. 8-21 Service head *(at right)* is for use when wires run through raceway. Use with special entrance ell *(at left)* which has a removable cover. *(Hubbell Electrical Products)*

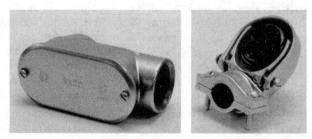

been listed for a conductor fill that meets your needs. For example, there are no "mogul" rigid nonmetallic conduit bodies, so (if listed to US standards) they are all routinely investigated for allowable wire fill. Look for a "3 4/0 XHHW MAX" or similar marking. If you aren't using XHHW wires, or if you're using combinations of sizes, just compare the actual cross-sectional area of the conductors with the area represented in the listing instructions. This fitting also must be raintight and arranged to drain; drill a small weep hole in its bottom end before you pull the wire.

After completing the pull, and assuming the raceway passes from the outdoors to an indoor location with air that can be significantly warmer or cooler than outdoor air, stuff some sealing putty around the wires where they enter the raceway. This will obstruct humid air from circulating to a point where condensation could occur at the colder end. The *NEC* has routinely enforced this principle on, for example, raceways passing from cold storage areas into warm rooms, and now expands on it, directly applying the principle to raceways (or cable sleeves) "passing from the interior to the exterior of a building."

Concentric knockouts   The cabinets of service switches or circuit breakers are too small to permit knockouts of all the sizes that might be required in all circumstances, so concentric knockouts are provided (Fig. 8–22). Remove the center section if you need the smallest size; remove the two smallest sections for the next size, and so on. Be very careful to remove only as many sections of this intricate knockout as required for your particular installation.

Fig. 8–22 Concentric knockouts are convenient, but great care must be used in removing each section.

MAKING GROUND CONNECTIONS

The service switch or circuit breaker has two heavy terminals to which the two incoming hot wires are connected. It also has a neutral busbar on which you will find several large connectors and a number of smaller terminals. Connect the incoming white or bare neutral to one of the large connectors, and the ground wire to the other. The smaller terminals are for the white wires of the individual circuits or feeders. If there are enough terminals, bare grounding wires of circuits can also be connected to the neutral bar *in service equipment only, never on the load side of this point.* If additional terminals are required, an accessory equipment grounding bar, available from the manufacturer, must be used. Although many terminal bars will be rated for multiple equipment grounding terminations in a single lug (often as many as three in the smaller wire sizes), never install more than a single grounded conductor (white wire) in one lug.

Solid neutral   Whether your service equipment contains breakers or fuses, the incoming neutral wire must never be interrupted by a breaker or fuse. Therefore in a three-wire, 120/240-volt installation, while there are three service-entrance wires, the main breaker is only a two-pole, and in fused equipment there are only two main fuses. The neutral is called a "solid neutral"—a neutral not protected by breaker or fuse. Such equipment then is called "three-wire SN."

Grounding   Every installation of the types discussed in this book must be grounded. The technical reasons for grounding are discussed in Chapter 7. A good, carefully installed ground is absolutely necessary for a safe installation. Grounding is accomplished by running a ground wire from the incoming neutral service wire (from a point where it is connected to the neutral busbar in the service equipment) either to an existing grounding electrode system or, if not available, to a grounding electrode specifically installed and connected for this purpose.

Grounding electrode system   All of the following that exist at the structure must be bonded together to form a grounding electrode system; in addition, any existing rod, pipe, or plate electrodes must be included with these:

■ Ten feet or more of buried metal water pipe.

■ The metal frame of a building where it has at least 10 ft of metal in direct earth contact, or where the metal frame is connected to a ground ring or concrete-encased electrode (See below).

■ A concrete-encased electrode consisting of 20 feet or more of ½-inch reinforcing steel or 4 AWG copper wire encased by at least 2 inches of concrete and located near the bottom of a concrete foundation or footing that is in direct contact with the earth. The requirement to connect to these electrodes can be waived, but only for existing buildings.

■ A ground ring of 2 AWG or larger bare copper wire encircling the building.

If only one of the above exists it may serve as the grounding electrode, *except* for water piping which *must* be supplemented by at least one other electrode. If none of these is available, use a rod, pipe, or plate electrode.

Rod, pipe, or plate electrodes, or other underground systems or structures   If there is no underground water system, a good ground becomes a problem. If none of the above options for a grounding electrode system is available, the *NEC* allows for several alternatives that may serve as the grounding electrode:

■ Underground metal piping (except gas) or tanks, including metal well casings.

■ Eight-foot driven ½-inch copper or ⅝-inch steel rod or ¾-inch galvanized steel pipe.

■ A metal plate buried at least 30 inches below the surface that exposes at least 2 square feet of surface to the soil. Since a plate has two sides, such plates are commonly 1 foot square in size.

Ground rod installation is discussed on pages 198–199. If a ground rod alone is used, the ground wire never needs to be larger than 6 AWG regardless of the size of the wires in your service equipment. The *NEC* requires rod, pipe, or plate electrodes to have a ground resistance or 25 ohms or less, if only one is installed. If the resistance rule cannot be met with a single such electrode, then an additional electrode must be installed. When more than one electrode is installed, the *NEC* effectively declares that you have tried hard enough, and does not insist on meeting the 25-ohm limit. There are many areas of the country where soil conditions routinely produce resistances in the hundreds of ohms. Check with local authorities to get a sense of local conditions. Many installers routinely install two such electrodes as a matter of course, avoiding the need for measurements. Be sure they are at least the *NEC* minimum of 6 feet apart, and even farther if possible.

Bonding the busbar   Service equipment cabinets contain a neutral busbar on which are installed several large solderless connectors for the heavy incoming neutral wire and the ground wire, plus as many smaller connectors or terminal screws as needed, one for the grounded wire of each 120-volt circuit. As purchased, the neutral busbar may be insulated from the cabinet. If you use this equipment as service equipment you *must* bond the neutral busbar to the cabinet. It must *not* be bonded if the breaker or switch is used, for example, as a disconnecting means for an appliance or other load, or is used as a panelboard on the load side of a service. If you have to bond the busbar to the cabinet, it is a simple matter. In some brands of equipment the neutral busbar is provided with a green bonding screw that you must tighten securely. In other brands, the cabinet will contain a flexible metal strap that is already bonded to the cabinet, and you must connect its free end to one of the connectors or screws on the neutral busbar. The equipment grounding busbar will always be bonded to the enclosure, whether within the service equipment enclosure or at a remote panelboard.

Installing the ground wire   There is usually only one ground wire from the service to the grounding electrode(s). There is no objection to using insulated wire, but bare wire costs less and is usually used for the ground wire. Keep the ground wire as short as possible. All grounding connections to the water pipe must be made within the first five feet of where the water pipe enters the building.

Size   The minimum ground wire size is based on the size of your service wiring. Larger size ground wire is optional. With 2 AWG or smaller service wires, 8 AWG is permitted for the ground wire but it must be enclosed in armor or conduit. It is more economical and practical to use 6 AWG because it doesn't have to be so heavily protected.

With 1 or 1/0 AWG service wires, 6 AWG is the minimum ground wire size. With 2/0 or 3/0 AWG service wires, 4 AWG is the minimum ground wire size. If you are using aluminum service wires, the equivalent cut points are 1/0 AWG or smaller, 2/0 or 3/0 AWG, and 4/0 AWG or 250 kcmil. Aluminum ground wires are also permitted. The *NEC* sizing rules extend to all service sizes, as covered in *NEC* Table 250.66.

Armor or conduit is not required for 6 AWG or 4 AWG ground wire. Either size must be fastened with staples to the surface over which it runs. It must be installed in a way that minimizes danger of physical damage after installation. Physical protection for the ground wire can also be provided by rigid nonmetallic (PVC) conduit.

Connecting   At the service equipment, connect the ground wire to the neutral busbar to which you have connected the neutral of the service. Run it to the grounding electrode(s), and there connect it to a ground clamp of the general style shown in Fig. 8–23. This clamp must be used on the size pipe and with the size ground wire marked on the clamp. If used with copper water tubing, it must be marked for this service; if used on reinforcing steel, it must be marked with the size.

Bonding at the service entrance   A careful job of bonding is essential to assure effective grounding. *Caution:* When using either steel armor or steel raceway to protect the ground wire, great care must be taken to securely and permanently bond the armor or conduit at both ends—to the cabinet and to the ground clamp at the electrode. Unless this is done, the resulting ground will be much less effective than when using unprotected wire. This means using a ground clamp that secures both

Fig. 8-23 Typical ground clamp for connecting ground wire to water pipe.

the raceway and the wire inside it simultaneously. At the service equipment, use a bonding procedure similar to that shown in *A* of Fig. 8–24 for the service conduit.

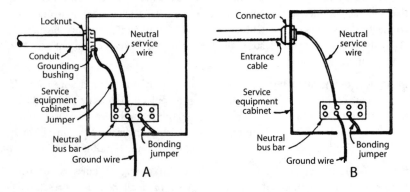

Fig. 8-24 Typical grounding connections. Only the white wire is shown. The wire marked "bonding jumper" is usually part of the service equipment as purchased; it could be either a green screw or a flexible strap.

Service-entrance wires   If your service-entrance wires are in conduit, follow *A* of Fig. 8–24. Install a grounding bushing (Fig. 8–25) on the end of the conduit inside the cabinet and run a jumper (bare or insulated, of the same size as the ground wire) from the bushing to the neutral busbar (grounding busbar, if downstream from the service). The screw of the bushing bites down into the metal of the cabinet, contributing toward a good continuous ground and preventing the bushing from turning and loosening, which

Fig. 8-25 Grounding bushing.

would impair the continuity of the ground. If your service uses service-entrance cable with its bare neutral wrapped around the insulated wires, as in Fig. 8–12, the grounding bushing is not required; follow *B* of Fig. 8–24.

When you use a threaded hub bolted to the enclosure, or a hub of the two-piece type shown in Fig. 8–26, bonding of the conduit to the enclosure is automatic and the grounding bushing and jumper are not necessary.

Ground wire   If you use armored ground wire, bond the connector on the cable to the cabinet, and at the electrode use a ground clamp similar to the type shown in Fig. 8–23 except with cable armor termination provisions; connect the wire to the terminal screw, and clamp the armor in the separate clamping device of the ground clamp.

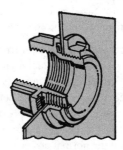

If the ground wire is protected by conduit, bond the conduit to the cabinet with a locknut and grounding bushing. A jumper from that bushing to the neutral busbar is required, as shown for the service conduit in *A* of Fig. 8–24. Use a ground clamp into which you can thread the conduit, and connect the ground wire to a terminal screw of the clamp. If the ground wire is protected by rigid nonmetallic (PVC) conduit, bonding of the raceway is not necessary.

Fig. 8–26 Two-piece threaded hub. Serrations on both pieces bite into enclosure metal, making a bonding jumper unnecessary. O-ring in outside piece forms a raintight seal.

Water meter   A bonding jumper must be installed around the water meter as shown in Fig. 8–27. Use two ground clamps and a length of wire of the size used for the ground. This prevents the ground from being made ineffective if the water meter is removed. Some water meters also have insulating joints. Use the same procedure around other insulated interruptions in the continuity of the water piping system, such as a water filter with plastic housings, even if the grounding connection is on the street side of such interruptions. Water piping systems must be bonded to the ground wire, and this often means looking in both directions from the ground clamp in order to find discontinuities.

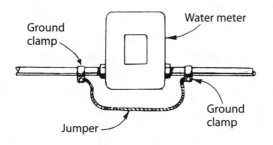

Ground clamp

Water meter

Jumper

Ground clamp

Fig. 8–27 Install a jumper around the water meter.

STANDBY BACKUP GENERATORS

There may be times when the main power supply is cut off due to a storm or other factors. If you live far from town, or in a lightning-prone area, your electric service may be subject to occasional or frequent outages. Experience will tell you whether a generator would be a good investment for you. Choice of a portable or permanently installed type will depend on your particular needs. Fuel types include gasoline, natural gas, diesel fuel, or liquid propane. Be sure to check local requirements for storage of fuel and the noise level permitted.

Portable or permanently installed?   Small portable generators can be used to power a few individual appliances such as the refrigerator, freezer, and a lamp or two. Be sure the extension cords you use to connect the appliances to the generator are sized to carry the specific loads. Check the watts/volts ratings on the nameplates of the appliances and on the extension cords. For convenience you may want a permanently installed generator. There are systems that automatically start the generator when the normal power source fails, and automatically transfer the load(s), but for the ordinary residential situation a system for manually starting the generator and transferring the load(s) will usually suffice.

Choosing the correct size   Generators are rated and priced according to the wattage they produce. Noise level and rate of fuel consumption (running time) are other factors to consider in making a selection. The size you need depends on whether you want to have the load of the entire house or farm on during an outage, or only a few selected circuits. A limited amount of rewiring might be required in order to separate the circuits for your essential equipment from the rest of the load, or in the case of a farm, the selected loads from the total load. If in doubt about how to proceed, consult with a professional.

See pages 47–50 and 190–192 for information on the calculation of residential and farm loads. For motors, multiply the horsepower by 746 to convert to watts. Use a generator with 25 percent capacity above the proposed connected load to permit motors to start. If your load includes hard-to-start motors, such as a well pump, it would be prudent to size the generator capacity to as much as three times the connected load in order to permit the motor to start. If some of the load is 240-volt, be sure the generator you choose will supply 240 volts, and not 120 volts only. A generator that is automatically started must be sized large enough to carry the entire load connected to it, startup as well as continuing. With a manually started generator you can turn off all equipment before starting the generator, and then restart selected equipment individually to minimize the load.

Isolate generator from main power supply   It is absolutely necessary that it not be possible for your generator and the incoming lines from the power supplier to be connected together. The principal reason for making sure your generator is isolated from the power supplier's lines is to protect the workers who are out on the power supplier's system trying to restore power—they could receive a serious shock or be electrocuted by current coming from your generator. Never attempt to energize

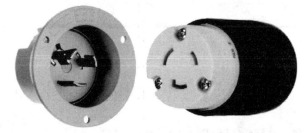

Fig. 8-28 Flanged inlet and the mating receptacle cord body that could energize it. *(Pass & Seymour/LeGrand)*

wiring in a building by using an extension cord with two male plug ends, or you will potentially create this and a number of other problems. The *NEC* requires that a sign be placed at the service location advising of the presence of the standby source.

Transfer switches are required to ensure this separation. If the generator neutral is bonded to its frame, the transfer switch should include an additional pole to switch the neutral. Transfer switches feed a given load from either of two alternate sources. Although breakers can be turned on and off to make load selections, Murphy's Law holds that the person available when the generator is required will be someone unqualified to make those decisions. At least post instructions for such an eventuality. A better procedure involves the use of an additional panelboard containing the circuits that are to be kept live during the period when the local power supplier is experiencing an outage, and feeding this panelboard from the transfer switch. Then connect a feeder from the usual service panelboard to one side of the transfer switch, and the line that will connect to the generator on the other side.

If the generator is portable and not permanently connected, use a plug-and-receptacle type of connection so you can make the connection easily and quickly. However, in this case the receptacle body must go on the end of the cord from the generator. This means the "plug" (the device with the male blades) must be at the building wall. These devices are called "flanged inlets" and although recessed into a wall, they have male blades, as shown in Fig. 8-28. This arrangement prevents plug blades from being fully exposed and energized while the generator is running.

Grounding the generator   The generator frame must be grounded to the same electrode as that used for the service. If the generator is not permanently installed, an equipment grounding conductor should run with the circuit conductors in the connecting cord. Be sure that after the transfer to the generator the identified (white, neutral) wire is still bonded to the grounding electrode conductor or system.

Chapter 9
OUTLET AND SWITCH BOXES

IN ORDER TO LIMIT SHOCK and fire hazard from poorly made or deteriorated connections and splices in wires, and to protect the wiring from physical damage, the *National Electrical Code* (*NEC*) requires that every switch, every outlet, and every joint in wire or cable must be enclosed in a box, and every fixture must be mounted on a box (with the exception of some fluorescent fixtures described on page 173). Keep in mind that all explanations in Chapters 9 through 15, except Chapter 13, are for "new work"—the wiring of a building while it is being built. The basic principles for "old work"—the wiring of a building after it is built—are covered in Chapter 13. Study new work thoroughly in order to better understand old work.

PLANNING YOUR BOX INSTALLATION

Many boxes are made of steel with a galvanized finish. Boxes made of insulating materials such as PVC, fiberglass, and other nonmetallic material are also common. Such boxes are used mainly with nonmetallic-sheathed cable and nonmetallic raceways. Metal boxes and their accessories are covered first because they are easier to visualize as we cover certain terminology common to all boxes.

Types of boxes Boxes are called switch boxes or outlet boxes depending on their shape and purpose. Wires and cables must be brought into the boxes through openings of the appropriate size, and "knockouts" are provided for this purpose. These are sections of partially punched out metal that are easily removed to form openings by loosening the knockout with a sharp blow on a screwdriver placed against it and then removing with pliers.

   Switch boxes The most common type of switch box is shown in Fig. 9–1. The mounting brackets on the ends of each box can be adjusted to bring the front of the box flush with the surface of the wall. Two "ears" with tapped holes for screws hold the switches, receptacles,

Fig. 9–1 A typical switch box used to house switches, receptacles, etc. The sides are removable.

and similar devices installed in the boxes. Switch boxes are sometimes used for fixtures of the wall-bracket type weighing up to 6 pounds, with a box located at each point where a bracket is to be installed.

The *NEC* requires boxes to have a depth of at least $^{15}\!/_{16}$-inch where a device such as a switch or receptacle is to be installed. Some are at least 1½ inches deep, but the deeper ones (2 to 3½ inches deep) are handier and more generally used. Boxes for large, flush-mounted equipment must be at least the depth of the equipment plus ¼ in.

Each box holds one strap. When two straps are to be mounted side by side, two boxes can be changed into a single "two gang" box by simply throwing away one side of each box and bolting the boxes together as in Fig. 9–2. Use the screws that are part of the boxes.

Fig. 9–2 Two single boxes are easily changed into one larger "two-gang" box. Made the same way are still larger boxes of three or four or more units ganged together. Ready-made two-gang and larger boxes are also available.

Outlet boxes   Figure 9–3 shows the most common outlet box. It is octagonal in shape and available in three sizes: 3 1/4-inch, 3 1/2-inch, and 4-inch. The 4-inch size is the most common today, permits more wires to be used, avoids cramping, and in general reduces the time required for installation. If space permits, use a box at least 1 1/2 or even 2 1/8 inches deep.

Outlet boxes must always be covered. A great many different covers are available. Some common ones are shown in Fig. 9–4.

Fig. 9–3 Typical octagonal and square outlet boxes.

Handy boxes   If an ordinary switch box is mounted on the surface of a wall, as in a basement or garage, the sharp corners of the box and cover are a nuisance. In such locations use a box with rounded corners, known as a "handy box" or "utility box" and shown in Fig. 9–5.

Square boxes   Boxes 4 inches square are large enough to hold two devices. Figure 9–6 shows a square box and the special covers used with it. These boxes are used for surface mounting as well as for general-purpose wiring and are especially handy when many wires must enter the same box.

How to calculate correct box size
NEC 314.16(A) specifies the maximum number of wires of the same size permitted in a metal box. Do not crowd a box to its limit. Doing so makes the work more difficult and time-consuming, and it runs a risk of shorts and grounds. Table 9–1 shows the interior cubic inch capacities of the more common standard metal boxes as taken from the complete list found in *NEC* Table 314.16(A). Nonstandard and nonmetallic boxes have their cubic inch capacities marked on them.

The total volume of an enclosure includes plaster rings, domed covers, extension rings, etc., which are marked with their volume in cubic inches. The space required for each conductor is shown in Table 9–2.

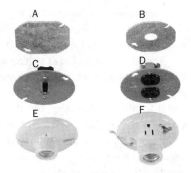

Fig. 9–4 Every outlet box must be covered. Some common covers: *A)* blank cover; *B)* knockout cover (will accept wiring entries); *C)* switch cover; *D)* cover with duplex receptacle; *E)* keyless receptacle for lamp; *F)* same as *E* but with pull chain and receptacle.

To find the minimum size box that is permitted, add up the required volumes listed in the box below. Choose all volumes from Table 9–2. Then match the total calculated volume in cubic inches to a box size in Table 9–1.

Example of box size calculation   Consider a feed-through receptacle location with two 12-2 with ground cables entering the box through internal cable clamps.

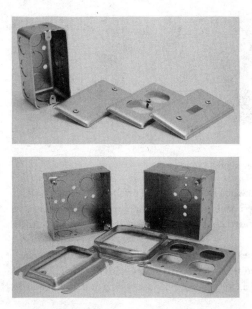

Fig. 9–5 This box is known as a "handy box" or "utility box." Use it for permanently exposed surface wiring as in basements, garages, etc. The box and covers have rounded corners for safety. *(Hubbell Electrical Products)*

Fig. 9–6 Four-inch square boxes and covers. Use covers on the left and middle when wiring is concealed. For permanently exposed wiring, use the type of cover shown on the right. *(Hubbell Electrical Products)*

From Table 9–2, each 12 AWG wire requires 2.25 cubic inches. The grounding conductors are also 12 AWG, so the calculations are:

four 12 AWG circuit wires	4 × 2.25 cubic inches = 9.00 cubic inches
two 12 AWG ground wires (count as one)	1 × 2.25 cubic inches = 2.25 cubic inches
two internal clamps (count as one)	1 × 2.25 cubic inches = 2.25 cubic inches
Duplex receptacle (count as two)	2 × 2.25 cubic inches = 4.50 cubic inches
	Total required 18.00 cubic inches

Thus from Table 9–1, the smallest square outlet box permitted is a 4 × 1¼-inch box, and the smallest switch box permitted for this application is a 3 × 2 × 3½-inch box. Both are 18 cubic inches.

INSTALLING BOXES

NEC 210.70 requires the following areas to be equipped with at least some lighting controlled by a switch: every livable room; bathrooms, hallways, stairways, attached garages; detached garages if wired; and the following spaces if used for storage or containing equipment requiring servicing—attics, underfloor spaces, utility rooms, basements. In addition, each entrance into the house must have an outdoor light controlled by a switch inside the house. All required lighting must consist of permanently installed lighting fixtures in bathrooms and kitchens, but in other habitable rooms it may be provided by floor or table lamps plugged into receptacles controlled by a wall switch.

Boxes for switches should be located near doors so the switches are easily accessible as the door is opened. Consider which way the door is to swing and install the box on the side opposite the hinges, with the center of the box 48 to 52 inches above the floor. For receptacles, the usual height is 12 to 18 inches above the floor. Each room should have at least one receptacle located conveniently for a vacuum cleaner and other portable items. Chapter 2 lists *NEC* minimums for horizontal spacing.

Mounting switch boxes   Boxes must be firmly secured to studs. Some boxes come with nails attached. If mounting nails pass through the box, they must not be more than ¼ inch from the back or ends. It is usually faster to use boxes with mounting brackets (Fig. 9–7). Always install boxes so their front edges are flush with the surface of the finished wall.

If the box is not adjacent to a stud, use an octagonal or square box with a plaster ring and a hanger (See next paragraph).

Mounting outlet boxes   If the box is to be installed near a stud, use an outlet box with a mounting bracket such as shown in Fig. 9–7. If the box must be mounted between studs

Fig. 9–7 Boxes with mounting brackets save time.

To calculate minimum box size, add up these required volumes, choosing volumes for wires from Table 9-2 below and then matching total calculated volume to a box size in Table 9-1.

1 VOLUME. . . . . . . for each wire originating outside the box and terminating or spliced within the box.

1 VOLUME. . . . . . . for each wire passing through the box without splice or termination (as may happen in conduit wiring).

1 VOLUME. . . . . . . for each fixture wire. Exception: four or fewer fixture wires and one equipment grounding wire entering the box from a domed fixture canopy are not counted.

1 VOLUME. . . . . . . for any number of internal cable clamps (Fig. 9-14), basing the volume on the largest wire in the box. Do not add for external cable connectors (Fig. 9-11).

1 VOLUME. . . . . . . for any number of fixture studs (Fig. 9-15 top), basing the volume on the largest wire.

1 VOLUME. . . . . . . for any number of hickeys (Fig. 9-15 bottom), basing the volume on the largest wire.

2 VOLUMES. . . . . . for each yoke or strap containing devices, etc., basing the volume on largest size wire connected yoke, and then doubling it. For a large device that won't fit in a single gang, two volumes for each gang required.

1 VOLUME. . . . . . . for any number of equipment grounding conductors, basing the volume on the largest equipment grounding conductor in the box.

Table 9-1	INTERIOR VOLUME OF STANDARD BOXES	
KIND OF BOX	SIZE IN INCHES	INTERIOR VOLUME IN CUBIC INCHES
Outlet box	4 × 1¼ ROUND	12.5
	4 × 1½ OR	15.5
	4 × 2⅛ OCTAGONAL	21.5
	4 × 1¼ SQUARE	18.0
	4 × 1½ SQUARE	21.0
	4 × 2⅛ SQUARE	30.3
	4¹¹⁄₁₆ × 1½ SQUARE	29.5
	4¹¹⁄₁₆ × 2⅛ SQUARE	42.0
Switch box	3 × 2 × 1½	7.5
	3 × 2 × 2	10.0
	3 × 2 × 2¼	10.5
	3 × 2 × 2½	12.5
	3 × 2 × 2¾	14.0
	3 × 2 × 3½	18.0
Handy box	4 × 2⅛ × 1½	10.3
	4 × 2⅛ × 1⅞	13.0
	4 × 2⅛ × 2⅛	14.5

Table 9-2	VOLUME REQUIRED PER CONDUCTOR
WIRE SIZE	VOLUME IN CUBIC INCHES PER WIRE
18	1.50
16	1.75
14	2.00
12	2.25
10	2.50
8	3.00
6	5.00

or joists, use a hanger of the type shown in Fig. 9-8. It is adjustable in length and has a fixture stud sliding on it. Remove the center knockout in the box,

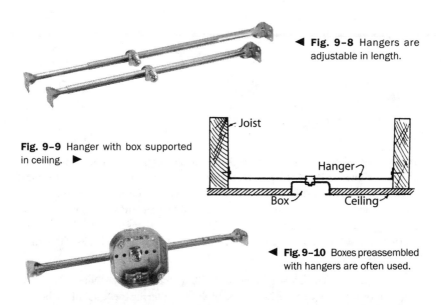

◀ **Fig. 9-8** Hangers are adjustable in length.

Fig. 9-9 Hanger with box supported in ceiling. ▶

◀ **Fig. 9-10** Boxes preassembled with hangers are often used.

slide the stud to the position wanted, push the stud through the knockout, and drive the locknut on the inside of the box securely home. Then nail the hanger to the studs or joists so the front edge of the box (or the cover mounted on it if it is the type shown in Fig. 9–6) will be flush with the wall or ceiling as shown in Fig. 9–9. Even handier are preassembled boxes with hangers as shown in Fig. 9–10.

Wiring at boxes   Regardless of the wiring method you use, let 6 to 10 inches of wire extend from the face of each box. The length depends on your choice of two methods of connecting the wire to the terminal screw as discussed on pages 37–40 and shown in Figs. 4–13 and 4–14. The *NEC* minimum has two aspects. The wires (including equipment grounding wires, as with Type NM cable) must be at least 6 inches long measured from where the individual wires emerge from the raceway connector or cable sheath, and they must be long enough to extend at least 3 inches outside the box. The only exception (other than for very large boxes) is for wires that have no splices or connections, but merely pass through the box, usually from raceway to raceway. These wires have no minimum length. They are only counted once in box fill calculations unless the size of the loop in the box is longer than 12 inches (in which case they are counted twice for each such loop).

Anchoring connectors   When cable of any style is used for wiring, the *NEC* requires that it be securely anchored to each box that it enters. There are many kinds of connectors for this purpose; an assortment is shown in Fig. 9–11. The connector at *A* is used for ordinary purposes, the one at *B* for a sharp 90-degree turn, and the one at *C* when two pieces of cable must enter the same knockout. To anchor the cable, first remove the locknut from the connector. Attach the connector to the cable with the clamp screw as in

A   B   C

Fig. 9-11 Connectors used in anchoring cable to boxes.

Fig. 9–12, then slip the connector through the knockout, and install the locknut inside the box as shown in Fig. 9–13. Be sure to drive the locknut down solidly so the lugs on it actually bite into the metal of the box to form a good continuous ground. Some newer connector designs snap into and hold tight to standard knockout openings, requiring no locknut. Some boxes have built-in clamps that securely hold the cable entering the box, making separate connectors unnecessary. Typical boxes of this kind are shown in Fig. 9–14.

Fixture studs   Small, lightweight (not over 6 lb) fixtures are often mounted directly on wall-mounted device boxes, anchored by screws entering the "ears" of the box as explained at the beginning of this chapter under the heading "Switch boxes." Other fixtures commonly mount to fixture bars that are secured to the box ears. Heavier

Fig. 9-12 Connector is first attached to the cable by means of a clamp screw.

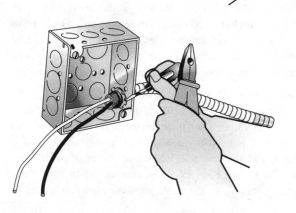

Fig. 9-13 Connector is then anchored to the box by means of a locknut inside the box.

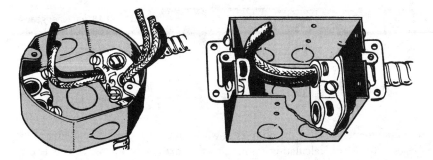

Fig. 9-14 With boxes of this type, separate connectors are not used. The boxes have clamps to hold the cable.

fixtures need additional support. Install a fixture stud (See Fig. 9–15) in the back of the box using stove bolts. This is unnecessary if a hanger (Fig. 9–8) is used to support a box mounted between studs or joists. With a fixture stud, a hickey may be needed between the stud and fixture. Fixtures above 50 pounds (or heavier if the box is so listed) must be supported directly by the building structure.

Boxes for suspended ceiling (paddle) fans   Boxes for the support of ceiling fans must be listed for the purpose. Fans weighing more than 35 pounds must be supported independently of the box, unless the box has been listed for additional weight. Consider using a box listed for fan support at every ceiling outlet location where a paddle fan might be installed at a later date, because installing a fan support box at an existing outlet can be difficult.

Fig. 9-15 Fixture stud and hickey.

Boxes for recessed fixtures   If the bottom of your fixture will be flush with the ceiling it is called a "recessed fixture." If your recessed fixture includes a junction box marked for 60°C branch circuit wiring, ordinary branch circuit wiring can be connected directly to it, and most recessed fixtures available have such boxes. Be sure to observe any fill or other restrictions on the number and type of wires allowed in such boxes, because they will not be of any standard size. See page 174 in Chapter 15 before installing.

Boxes for interchangeable devices   Up to three interchangeable devices can be installed in an ordinary single-gang box. Figure 9–16 shows a faceplate with three openings, and an assortment of devices such as switches, receptacles, and pilot lights. Plates are also available with one or two openings. The plate is supplied with a metal strap with three openings on which you can mount any combination of one, two, or three of these specially sized devices. The three devices can then be installed in a single-gang box. Three standard devices would require a three-gang box which costs more, takes more space and requires much more installation time.

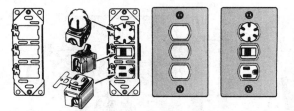

Fig. 9-16 Interchangeable devices permit 3 switches, receptacles, and the like to be installed in a single-gang box. A common disconnect is required—see pages 116-117.

Caution: Remember that the number of wires entering any box is limited; review pages 100–101 on calculating minimum size boxes. When a switch box is not large enough for the wires necessary to connect to three devices on one strap, use a 4-inch-square box and a single-gang plaster ring as shown in Fig. 9-6.

Nonmetallic outlet boxes   Because of lower cost, nonmetallic boxes have largely supplanted metal boxes, particularly for residential occupancies. In addition, nonmetallic boxes are largely immune to corrosion, and, for surface applications, using boxes of PVC, fiberglass, or other nonmetallic material provides insurance against shock. Metal boxes used in barns and similar locations can rust out just like conduit, and if their grounding connections are compromised, they can easily present a shock hazard if there is an insulation failure on any hot wire inside.

Fig. 9-17 Especially on farms, boxes made of insulating material are frequently used.

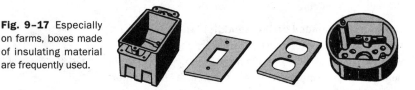

Nonmetallic boxes (and covers) are shown in Fig. 9-17. Use them as you would metal boxes; however, the *NEC* does not require that cable be anchored with connectors to single-gang nonmetallic switch boxes if the cable is supported within 8 inches of the box measured along the cable sheath, and the sheath extends into the box at least ¼ inch. At other than single-gang nonmetallic boxes, the cable must be secured to the box, as it must be to all metal boxes. (Anchoring with connectors is discussed on pages 103–104.)

Other nonmetallic boxes are shown in Fig. 9-18. Boxes with hubs, like that on the left, are for surface wiring, and with the proper covers they are weatherproof. Boxes with knockouts can be used either concealed or exposed. Unlike metallic boxes, nonmetallic boxes do not come in standardized sizes. Thus in order to decide on allow-

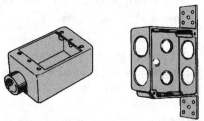

Fig. 9-18 PVC boxes with integral hubs and with standard knockouts.

able wire fill, you have to do the calculations covered earlier in this chapter, based on the actual volume of each box, which will be marked on the inside of the box.

Nonmetallic faceplates or covers are recommended when using nonmetallic boxes, particularly in corrosive locations. If metal faceplates are used, they must be grounded, which is more or less automatic with receptacles, because the mounting strap is usually connected to the green grounding terminal. Toggle switches require a special switch having a grounding terminal on the mounting strap to which the grounding wire in the nonmetallic-sheathed cable must be connected.

SELECTING SWITCHES

While all common switches have the general appearance shown in Fig. 2–1, there are three varieties that you should understand:

Single-pole is used to turn a light on or off from one point. It has two terminals and the words ON and OFF on the handle.

Three-way is used to turn a light on or off from two separate locations. It has three terminals and a plain, unlabeled handle. For the wiring of three-way switches, see pages 58–61.

Four-way is used in combination with three-way when a light must be turned on or off from more than two locations. Use three-way switches at two of the points, and four-way at the remaining points. The four-way has four terminals and a plain, unlabeled handle. For the wiring of four-way switches, see pages 61–62.

Switch ratings   Switches are designed for use at an amperage and voltage not higher than the limits stamped into their metal mounting yokes. Some switches are rated "10A 125V–5A 250V" indicating that the switch may be used to control loads not over 10 amps if the voltage is not over 125 volts, but only 5 amps if the voltage is higher but not over 250 volts. Most are rated 15A 120V, or 15A 120/277V. Other switches are available for higher amperages.

Grounding for switches   The yokes of all toggle switches, including dimmers as described below, are required to be grounded so that a metal faceplate, whether installed or not, will be grounded. The switch yoke can be grounded to the grounded metal box, or an equipment grounding conductor must be connected to a green terminal screw on the yoke.

Types of switches   In addition to the common general-use switches, there are switches with a variety of special features that enhance comfort, convenience, and safety. Some of these are described here.

AC general-use snap switches   The most commonly used switch today is called the "ac general-use snap" type in the *NEC*. It is very quiet in operation and lasts a long time. It is identified by the letters "AC" following the ampere and voltage rating on the yoke. As the name implies, it must not be used on dc circuits. On ac, it may be used for any purpose with two exceptions: a) it must not be used to control tungsten-filament lamps (ordinary lamps) at voltages above 120 volts; b) if used to control a motor, it may be used only up to 80 percent of its ampere rating.

AC-DC general-use snap switches   Older style switches are called "ac-dc general use" in the *NEC*. If the switch does not have "AC" at the end of its rating, it is an ac-dc type. This style may be used on dc-only circuits as well as on ac circuits. If your wiring was installed many years ago, the switches are likely to be the ac-dc type. If one of them fails, you may replace it with the more readily available ac-only type.

Lighted switches   Switches are available in two forms with tiny lights in their handles. The "lit handle" is on when the load is off to make it easy to locate the switch in the dark. The "pilot handle" version lights when the load is energized, so you can remember to turn it off when done. The lights consume so little power that they cost perhaps one cent per year to operate.

Quiet switches   Ordinary switches can make an audible click when turned on and off. The "ac general use" switch is very quiet. There is also a completely silent switch that uses a bit of mercury in a glass tube to make and break contact. Mercury switches operate only when installed in a vertical position. They are disappearing from the market due to environmental concerns about mercury disposal.

Dimmer switches   Where it might be desirable to control the level of brightness, such as in dining rooms or family rooms, ordinary switches can be removed and replaced with special dimming switches. Some dimmers are designed to be used only with incandescent lamps and others are for use only with fluorescent lighting.

One inexpensive type of dimmer switch provides HIGH–OFF–LOW positions and controls up to 300 watts. It can be used to replace only an ordinary single-pole switch, not a three-way. The somewhat more expensive type controls brightness continuously from off to full brightness. Some models are available in the three-way type, replacing one of a pair of three-way switches.

Fig. 9–19 Every switch and receptacle must be covered with a faceplate.

FACEPLATES

Receptacles and switches when installed are covered with faceplates (also called wall plates) of the type shown in Fig. 9–19. These are available in molded plastic and nylon in an assortment of colors and in other materials such as steel, brass, or ceramic. Use the kind you like best. Metal faceplates must be grounded.

Chapter 10
BASIC WIRING PROCEDURES

IN RESIDENTIAL AND FARM WIRING, the great majority of new wiring involves the use of cabled methods (covered in Chapter 11), among which the use of nonmetallic-sheathed cable predominates. It is the least expensive in terms of material cost, and requires the least technical training to install. However, armored cable and metal-clad cable are widely used, particularly with metal framing members. Tubular raceway methods (covered in Chapter 12) are less common but still used because the wires within them may, within the raceway size limitations, be replaced with different wires to serve different needs. The knob-and-tube wiring system, not permitted for new work since the 1975 *NEC*, is only for extensions in existing work (See pages 146–147).

Local codes sometimes prohibit one or more of these systems. What is prohibited in one locality may be required in another. For farm buildings, some form of nonmetallic-sheathed cable is used almost exclusively. It is also used extensively in residences and larger buildings, though highrise construction often uses metallic methods. Follow local custom. If you are not sure, consult qualified local authorities.

Many wiring procedures are the same regardless of which system is used. These procedures are explained in this chapter. It is essential to understand them thoroughly. The diagrams and explanations in this chapter use cable as a model, but the principles apply generally to all wiring systems. Details specific to cable or raceway methods are described in Chapters 11 and 12.

PLANNING THE INSTALLATION
Before doing any actual wiring, you must make a plan. Decide upon the location of each outlet and each switch. Be generous in the number of outlets and switches. Review Chapter 2 on planning an adequate installation. Remember that adding an outlet later costs much more than including it in the original job.

Which outlets on which circuit?   Your first idea may be to put all your basement outlets on one circuit, all first-floor outlets on the next circuit, all second-floor outlets on still another circuit, and so on. If you do that and a breaker trips or a fuse blows, an entire floor will be dark. Instead, put different parts of any floor on different circuits. Then at least part of each floor will still be lighted if any circuit goes out. One of the house circuits can be extended to an attached or separate garage that has small power requirements, or separate circuits can be used for a garage and accessory buildings requiring power (greenhouse, studio, machine shop, etc.). Wiring for garages and accessory buildings is discussed at the end of this chapter.

Diagram your circuits   After you have decided which outlets are to go on each circuit and where you want single-pole switches and three-way switches, draw a diagram of each circuit showing how the cable is to run from one outlet to the next, from outlet to switches, and so on. Figure 10–1 is a typical diagram for an installation using cable. It shows eight outlets, plus three single-pole switches and a pair of three-way switches. The outlets are labeled *A, B, C, D, E, F, G,* and *H.* Switches have all been labeled *S.* The switch controlling outlet *B* is indicated as *S-B,* the one controlling outlet *G* as *S-G,* the one controlling outlet *H* as *S-H,* and the two 3-way switches controlling outlet *F* as *S-F-1* and *S-F-2.* (Outlet A is a light controlled by a pull chain and is included here as part of the general explanation. Note that pull chains are not recommended; switches are preferred for convenience and safety.)

This diagram still does not tell you how to connect the individual cable wires so that all the parts will work properly. Draw a second diagram showing the wires separately as if they were open wires instead of in cable. In order to simplify the diagrams and explanations, the cable's bare grounding wire is omitted here, but requirements and instructions for connecting it are given on pages 128–129 in Chapter 11 under the heading "Use cable with grounding wire." Figure 10–2 shows the same outlets as Fig. 10–1, and each outlet has been labeled the same as in

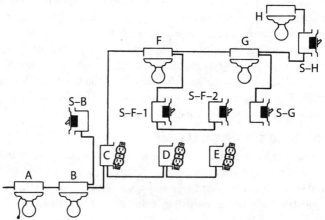

Fig. 10–1 Before doing any wiring, make a rough layout of each circuit showing location of each outlet and how cable will run from box to box. Shown here is a layout of a circuit with 8 outlets labeled *A* to *H,* 3 single-pole switches *S-B, S-G,* and *S-H,* and a pair of 3-way switches *S-F-1* and *S-F-2.* The wires for this circuit begin in the basement and run to main floor living room, kitchen, and dining room, and then to a bedroom.

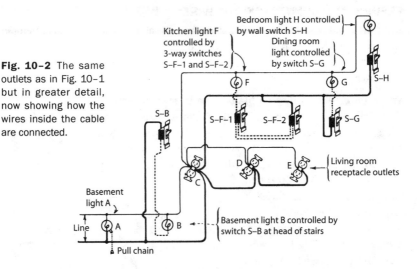

Fig. 10-2 The same outlets as in Fig. 10-1 but in greater detail, now showing how the wires inside the cable are connected.

Bedroom light H controlled by wall switch S–H

Kitchen light F controlled by 3-way switches S–F–1 and S–F–2

Dining room light controlled by switch S–G

S–H

S–B

S–F–1   S–F–2   S–G

Living room receptacle outlets

D   E

C

Basement light A

Line   A   B   Basement light B controlled by switch S–B at head of stairs

Pull chain

Fig. 10-1. Note that the same wire identification scheme has been used as in Chapter 6: a light line like this ——— for a white wire, a heavy line like this ▬▬ for a black wire (or other color, but not white or green), and a heavy broken line like this ▬ ▬ ▬ ▬ for the wire between a switch and the outlet it controls.

If you refer to the diagramming of basic circuits covered in Chapter 6, you will see that Fig. 6–3 is the same as outlet *A* in Fig. 10–2 (if you disregard the cable that runs on to *B*). Figure 10–3 shows this same outlet completely installed with an outlet box. Following the principles presented in Chapter 6, you run the white wire from SOURCE to the outlet, and you run the black wire from SOURCE to the outlet.

To better understand the wiring of each additional outlet, consider the cable that runs to it from the *previous* outlet as the source for the *new* outlet. For example, the cable that runs from outlet *A* to outlet *B* becomes the source for outlet *B*.

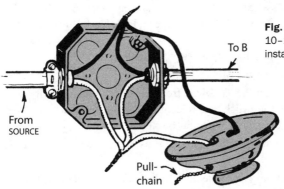

To B

Fig. 10-3 Outlet *A* of Figs. 10-1 and 10-2, completely installed.

From SOURCE

Pull-chain

FEEDING AND CONNECTING THE WIRES

If you are using nonmetallic-sheathed or armored cable, you will need to remember the important *NEC* requirement described here regarding wire color in making connections at switches. Following that, the basic guidelines for making connections apply to either cable or conduit.

White wire reidentified as black wire in switch loops   In Fig. 10–4, if you disregard the cable running to *C*, outlet *B* is the same as that in Fig. 6–5, so a detailed explanation may seem unnecessary. However, according to everything you have learned up to this point, and as shown in Fig. 10–2, *both* wires from outlet *B* to switch *S-B* should be *black*, but the two-wire cable that you are going to use contains one black and one white wire. How can you comply with the *NEC* requirement? *NEC* 200.7 permits a white wire to be used where a black wire *should* be used—but only in a switch loop (the cable running from an outlet to a switch), and the white wire must be reidentified as black wherever it is visible and accessible. This can be accomplished by painting, wrapping with black tape, or

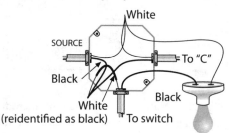

Fig. 10–4 Outlet *B* of Figs. 10–1 and 10–2. This is an important diagram. It shows how to connect white wire in cable to the switch.

sleeving the wire with a tube of either black plastic insulation removed from a larger wire or with plastic shrink tubing applied with a heat gun. Regardless of whether the white wire is reidentified or not (the *NEC* prior to 1999 did not require this), it must be used as the wire running to the switch from the ungrounded conductors in the outlet, never the reverse. That means the connections at the outlet will always be white to white and black to naturally black, avoiding the hazard of reverse polarity at a screw shell or receptacle. See Fig. 10–6 for an example.

Wiring at outlets and switches   It is easy to make the right connections if you remember that each fixture must have one white and one black wire connected to it, and if you observe the following simple steps:

1. At the switch, connect the two wires of the cable to the switch.

2. At the outlet, connect the white wire from SOURCE to the fixture, as usual.

3. At the outlet, connect the *black* wire from SOURCE to the *white* wire of the cable that runs to the switch, and reidentify the white wire as black. This is the only case where a white wire may be connected to a black.

4. Connect the black wire of the cable that runs to the switch to the fixture, as usual.

5. The two wires running on to the next outlet are connected to the two incoming wires (from source) in the outlet box—black to black, and white to white.

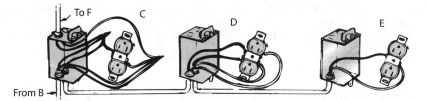

Fig. 10-5 Receptacle outlets are easy to connect, as this diagram shows.

When outlet *B* is properly installed according to these five steps, in accordance with *NEC* 200.7, it may yet fail 404.2(C); see page 63 and Fig. 10-6.

Outlets *C, D,* and *E* are wired as shown in Fig. 10–5. Receptacles have double terminal screws so that the two wires from two different pieces of cable can easily be attached as shown. In the case of *C*, it is necessary to connect three different wires to each side of the receptacle, but there are only two terminal screws. *NEC 110-14(A) prohibits more than one wire under one terminal screw.* Splice all the blacks together and all the whites together, adding a short piece of wire called a "pigtail" to the white set and another to the black set, and connect these pigtails to the receptacle terminal screws. See Fig. 4–15.

Outlet *F* in Fig. 10–2, if you disregard the cable running on to the next outlet *G*, is the same as Fig. 6–14 in Chapter 6. Wire it as shown in Fig. 10–6, which takes into account the new *NEC* requirement for neutrals run to most switching points. Run three-wire cable from the outlet to the first three-way switch *S-F-1* and four-wire cable from there to the second three-way switch *S-F-2*. Again you meet the problem of the proper colors of wire. Remembering the steps outlined in connection with outlet *B*, simply connect the white wire from source (which in this case is the white wire from outlet *C*) to the fixture as usual. Continue the white wire to both switch locations. The other wire on the fixture must not be white, so connect the red wire of the cable that runs on to the first switch *S-F-1*. The cable that runs on to the next outlet *G* must also be connected in any convenient box, black wire

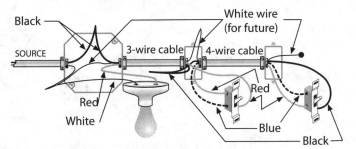

Fig. 10-6 The wiring of outlet *F* of Figs. 10–1 and 10–2, with the two 3-way switches that control the outlet. This is the diagram for any outlet controlled by 3-way switches when using cable. Study carefully the colors of the wires in this cable. Compare it with Fig. 6–14, and also page 63.

to black and white to white of the cable from SOURCE (from C). This completes the wiring of the outlet F. The switches still need to be connected.

Two different cables end in the box for S-F-1: a three-wire cable and a four-wire cable, making seven wires altogether. Two of them are white and two are black; splice each color together so there will then be a continuous white wire from F to S-F-1 to S-F-2 for the future, and a continuous black to feed S-F-2. Connect the red wire in the cable between F and S-F-1 to the common or marked terminal of the first switch S-F-1. That leaves two unconnected wires in the cable from S-F-1 and S-F-2: the red and the blue. Connect them to the remaining traveler terminals of each switch; it does not matter which color goes to which terminal on the switch. That finishes the wiring.

If you are installing four-way switches, follow the diagrams of Figs. 6–19 and 6–21. White wire may be reidentified where otherwise a wire of a different color would be required.

Outlet G is exactly the same as outlet B and should be wired in the same way.

Feed-through switch box   In all the outlets wired so far, you have run the cable first to the outlet box and fixture, then on to the switch. When you come to outlet H of Fig. 10–2, you will see that the cable runs first to the switch box S-H and then on to outlet H. This combination is even simpler to wire than the others because there is no problem with the colors of the wire, as you can see from Fig. 10–7 showing this outlet completely wired.

Splicing for additional outlets   You cannot install additional outlets beyond H by simply connecting the black and white wires of the cable for the new outlet to the black and white wires in outlet H, because then the new outlet would be turned on and off by switch S-H. However, you can add an additional outlet by tapping in at switch S-H; splice the wires of the cable for the new outlet to the incoming cable from G, as shown in Fig. 10–8. Another way is to run three-wire cable from S-H to H; splice the two wires in the cable for the new outlet to black and white in H, as shown in Fig. 10–9.

A cable to any additional outlet can be spliced to any existing cable in any outlet box by splicing white to white, and black to black—as long as each wire can be traced all the way

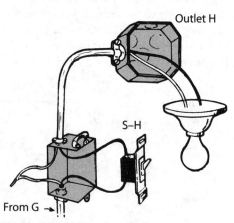

Outlet H

S-H

From G →

Fig. 10–7 Sometimes the cable from source does not run first to the outlet and then to the switch. Use this diagram when the cable runs first to the switch and then to the outlet.

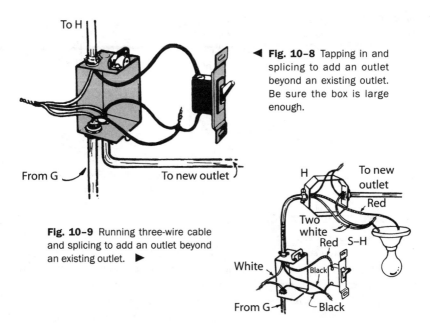

◀ **Fig. 10–8** Tapping in and splicing to add an outlet beyond an existing outlet. Be sure the box is large enough.

Fig. 10–9 Running three-wire cable and splicing to add an outlet beyond an existing outlet. ▶

back to SOURCE without interruption by a switch.

Three-way switches substituted for single-pole It is a simple matter to substitute two 3-way switches for a single-pole switch in any wiring diagram. Study Fig. 10–10: the starting point is an outlet already wired with two wires ready for a switch. If a single-pole switch is to be used, connect it to the two ends of the wires as at *A*. If three-way switches are to be used, substitute the combination of *B*; but keep track of the future neutral requirement.

Adding a switch to an unswitched outlet If you want to add a switch to a diagram that shows an outlet permanently connected without a switch, cut the black wire. That gives you two ends of black wire to which the switch connects, or two ends to which you will splice the two-wire cable that runs to the switch.

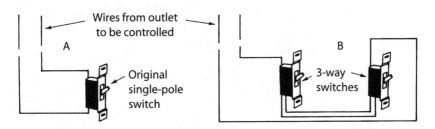

Fig. 10–10 It is a simple matter to substitute a pair of three-way switches for a single-pole switch in any wiring diagram. Simply substitute the right-hand combination for that at left.

Junction boxes Sometimes it is necessary to make a T connection in cable when there is a long run and no convenient outlet from which to start the T branch. Use an outlet box; run the three (or more) ends of cable into it; splice all black wires together and all white wires together; cover with a blank cover as shown in Fig. 9–4A, and the job is finished. A junction box is shown in Fig. 10–11. *Remember: Junction boxes must always be located where permanently accessible.*

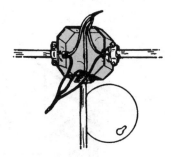

Fig. 10–11 A junction box contains only the splices of several lengths of cable.

TWO-CIRCUIT DUPLEX RECEPTACLES

It can be convenient to have a floor lamp or table lamp plugged into a receptacle controlled by a wall switch. These two-circuit duplex receptacles (also called split-wired receptacles) meet the *NEC* requirement for switch-controlled lighting in habitable rooms other than kitchens and bathrooms. Many brands of ordinary duplex receptacles are constructed so you can, during installation, change them to the two-circuit variety by breaking out a small brass portion between the two "hot" (brass colored) terminal screws. The wiring diagram is shown in Fig. 10–12. Actually, as the drawing makes clear, using a split receptacle in this way does not place it on two different circuits, because the same circuit breaker or fuse ultimately supplies both halves of the duplex receptacles shown. If that is not the case, *NEC* rules make the installation much more problematic.

The *NEC* now places a very significant condition on the use of multiple receptacles, or even other devices, such as a switch and receptacle combination, on one yoke (this is the technically correct way to describe a split duplex receptacle—the *NEC* counts each half of the duplex as a separate device) connected to more than one

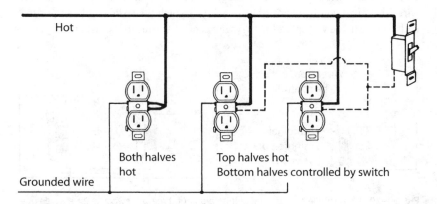

Fig. 10–12 Two-circuit receptacles are well worth the extra cost.

branch circuit. A means to simultaneously disconnect all ungrounded conductors arriving at the receptacle must be provided at the panelboard. Theoretically, this could be a multipole circuit breaker or a multipole switch installed adjacent to the panelboard. The *NEC* specifically allows such a snap switch as long as it is positioned "at the point where the branch circuits originate" and as long as the voltage rating of the switch at least equals that of the highest circuit voltage applied. If the two circuits are supplied from different line buses, that voltage will be 240V, and a lesser switch rating could not be applied. However, 277V snap switches are commonly available, and will work for this purpose. Such switches can also be used in remote locations, provided the common-disconnect rule is observed at the panel. In the case of a multiwire branch circuit, this common-disconnect rule is easily met with a two-pole circuit breaker, but if so wired it cancels one of the functional advantages of the traditional multiwire branch circuit, namely, if one circuit trips the other is still connected.

In addition, if two 2-wire circuits arrive at the device yoke, these circuits must still now originate from a multipole circuit breaker. In the case of the dining room split receptacle described previously, the multipole circuit breaker required here means the small-appliance branch circuit connected to this outlet could not be multiwired with a second small-appliance branch circuit, contrary to the usual configuration, unless the 2-pole snap switch at the panel is supplied. Think carefully about the implications of this rule in planning your installation. This rule applies in all occupancies, and any time more than one branch circuit supplies more than one device on a common yoke, whether or not in a multiwire configuration. The concern is that an untrained person will assume all circuits are off when investigating a nonfunctioning load connected to one of the receptacles.

SPECIAL RULES ON RECEPTACLE DESIGN

Tamper-resistant receptacles   Many toddlers are injured when they insert metal objects into energized receptacle slots in homes where the parent has failed to take elementary precautions to prevent this, such as by inserting inexpensive, commonly available insulating guards into unused receptacle slots. The *NEC* requires that all receptacles specified by 210.52 (the dwelling receptacle placement rules covered in Chapter 2) be listed as "tamper resistant." These receptacles have linked internal shutters that prevent access to energized parts unless both slots are engaged simultaneously; this rule now also applies to receptacles being replaced.

Weather-resistant receptacles   To address concerns about the stability of key receptacle components under exposure to weathering, the *NEC* requires all nonlocking receptacles rated 15A or 20A, 125V or 250V, if located in damp or wet locations, to be listed as weather resistant. This will be shown by a "WR" designation on the face. The requirement applies to both residential and nonresidential applications alike; this rule likewise now applies to receptacles being replaced.

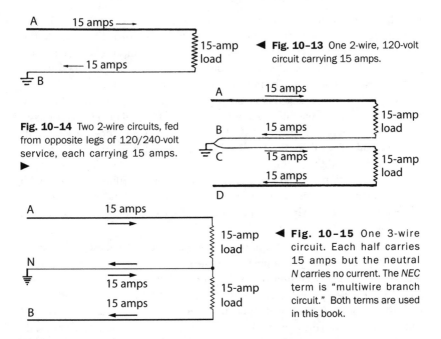

Fig. 10-13 One 2-wire, 120-volt circuit carrying 15 amps.

Fig. 10-14 Two 2-wire circuits, fed from opposite legs of 120/240-volt service, each carrying 15 amps. ▶

Fig. 10-15 One 3-wire circuit. Each half carries 15 amps but the neutral N carries no current. The NEC term is "multiwire branch circuit." Both terms are used in this book.

THREE-WIRE (MULTIWIRE) CIRCUITS

In terms of material cost, installation time, and voltage drop, installing one 3-wire circuit offers many benefits over having two 2-wire circuits.

An ordinary 2-wire, 120-volt circuit consists of two wires—the grounded wire and a hot wire as shown in Fig. 10-13. When two such 2-wire circuits are run to the same general area, there is a total of four wires as shown in Fig. 10-14 in which wires B and C are the grounded wires, and the wires A and D are hot wires *connected to opposite sides* of SOURCE, so that the voltage between A and D is 240 volts, but between A and B and also between B and C it is 120 volts.

But note that the two grounded wires B and C are connected together at the neutral busbar in your service equipment. Since they are connected together at the starting point, they become in effect one wire. Why use two? There is no need for two. Use one wire and instead of two 120-volt, 2-wire circuits you will have one 3-wire, 120/240-volt circuit as shown in Fig. 10-15. Be sure that each receptacle is connected to the neutral wire to obtain 120 volts. If connected to the two hot wires the voltage would be 240 volts.

Installing a 3-wire circuit not only saves time and material, but also reduces the voltage drop: if the loads on each half of the circuit (between A and BC, or between D and BC) are equal, the neutral wire carries no current at all.

Important: In Fig. 10-14, each wire B and C is merely a *grounded* wire, but in Fig. 10-15, the single wire N (formerly BC) will, if both loads are equal, carry no current at all, so it becomes a grounded *neutral* wire. If the loads are unequal, it will carry only the difference between the current in A and that in B, but it is still a neutral.

Where used   One 3-wire circuit is especially recommended in place of two 2-wire circuits for the small-appliance circuits discussed in Chapter 5. If the circuit is long, as for a yard light or as a feeder to a detached accessory building, voltage drop is a consideration (See page 29). Other loads commonly served by 3-wire circuits are the dishwasher and waste disposer, the microwave oven and kitchen sink instant hot water heater, and the range and dryer. Even other general-use circuits often start out at the panelboard as 3-wire circuits to the first one or two outlets, where they split off to become two 2-wire circuits.

Connecting at the service   Great care must be taken to connect wires A and B (the hot wires) to opposite legs of the incoming service-entrance wires. Normally the wire N carries only a small number of amperes, or no current at all if the loads between A and N, and B and N, happen to be identical. But if you connect A and B to the same leg of the incoming service wires, then wire N would be carrying a double load, possibly as much as 40 amps if 12 AWG wire is used; its insulation would be damaged and fire might result. Remember the grounded wire is not protected by a fuse or breaker. The *NEC* also requires that multiwire circuit conductors, including the neutral, be grouped in the originating panel with wire ties or equal, unless the circuit enters its own raceway or cable assembly that makes the grouping obvious.

Three-wire circuits at receptacle and lighting outlets   When adding receptacles to a three-wire circuit, the neutral must be especially watched. A receptacle has two terminal screws on each side. When wiring two-wire circuits, it is the usual practice to end two different grounded wires at these two terminals, as is done especially when wiring with cable. But in wiring three-wire circuits, this must not be done because removing a receptacle, such as for replacement, would then result in a break in the neutral wire during the time there is no connection to the receptacle. This would in turn place all the receptacles connected to one leg (beyond the receptacle temporarily removed) in series with those connected to the other leg, all at 240 volts. Appliances would malfunction and be ruined or at least badly damaged, and lamps would burn out.

Maintaining a grounded conductor continuity in three-wire circuits   If using conduit, run a continuous wire with a loop as shown in Fig. 4–17. When using cable, there is no way of making a loop so you must make a pigtail splice as shown in Fig. 4–15. Connect the ends of the white wires of two lengths of cable to each other and to another short piece of white wire, the opposite end of which you then connect to the white terminal screw of the receptacle. It is good practice to pigtail both the grounded and hot wires at all receptacles where you would otherwise use the two terminals on each side for feeding through. In this way the removal of one receptacle does not interrupt service to other receptacles downstream, nor do the downstream receptacles depend on the terminal screws or push-in connections as a substitute for a good splice. Pigtail neutrals at lighting outlets also, so that replacing a fixture will not open the neutral to downstream outlets.

Installing split-wired receptacles on three-wire circuits   The split-wired (also called two-circuit) receptacle shown in Fig. 10–12 can also be installed on the three-wire circuit, wired with one half on phase A, the other on phase B (Fig. 10–15), and a common neutral. This is often done in the kitchen, where two small-appliance receptacle circuits are required. Take care to observe the common-disconnect rule, or consider installing two duplex receptacles at one outlet, with the yokes connected to different circuits.

DETACHED GARAGES AND ACCESSORY BUILDINGS
Lighting and receptacle requirements for garages and accessory buildings are discussed on page 18 as well as in this section. The feeder from the house to the detached building may be run underground or overhead.

Running underground or overhead wiring   If running the feeder underground, you may use Type UF cable, which can be buried directly, or regular circuit wires with "W" in the name can be installed in a raceway such as rigid nonmetallic (PVC) conduit. The feeder must be buried at least 18 inches below grade if run in nonmetallic conduit; UF cable needs 24 inches of cover. Heavy-wall steel conduits require only 6 inches of cover. See further discussion in the chapter on farm wiring on pages 195–196.

If you are going to use overhead wiring, there are several ways that the wires can be brought out of the house and into a separate garage or accessory building. Figure 10–16 shows a very convenient entrance cap requiring only a single hole through the wall. The cap is used at both ends and is suitable for use with cable as shown. You can also use conduit by running it directly into the fitting, which accommodates ½-inch conduit.

Shown in Fig. 10–17 is another way of using an ordinary entrance cap (as used for service entrances), a short piece of conduit, and an outlet box inside the building.

Wiring for lighting and receptacles   For a light in a garage or accessory building that can be controlled both there and at the house, three wires need to be run from the house and a three-way switch installed at each end, wired as shown in the basic diagram of Fig. 10–18.

For receptacles that operate independently of the garage light so they are always live whether the light is on or off, four wires are required between house and garage. Follow the diagram of Fig. 10–19.

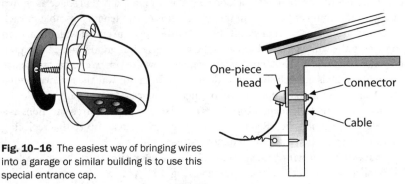

One-piece head

Connector

Cable

Fig. 10–16 The easiest way of bringing wires into a garage or similar building is to use this special entrance cap.

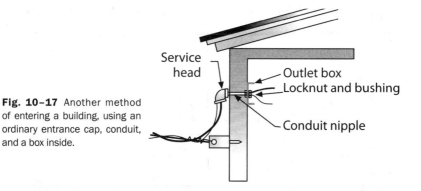

Fig. 10-17 Another method of entering a building, using an ordinary entrance cap, conduit, and a box inside.

Service head

Outlet box

Locknut and bushing

Conduit nipple

Calculating feeder wire size   Feeder wire size is based on the size of the load it must carry. To calculate the load of an accessory building that is a woodworking shop, machine shop, ceramic shop, greenhouse, etc., be sure to include the nameplate ratings of all the equipment. Provide plenty of light, calculated at not less than 3 watts per square foot, and add 180 volt-amperes for each general-use receptacle. Don't forget outside lighting so you can safely get back to the house after dark. If the accessory building is some distance from the house, calculate the voltage drop, or use Table 4–3, so the feeder will not be undersized.

Grounding   The former allowance to wire the feeder as if it were a service, with the neutral and grounding conductors bonded at the building disconnect and to the grounding electrode conductor at the second building, is severely limited. It cannot be used if there is any conductive path between buildings, such as a common metallic water piping system or the use of metal pipe as the feeder raceway, that could return neutral current over any route other than that of the white wire that should perform this function. It also applies to "installations made in compliance with previous editions of this *Code* that permitted such connection," so it would not be allowed in new construction.

The preferred method uses a separate equipment grounding bar along with a separate equipment grounding conductor that is run with the feeder conductors to the detached building. A grounding electrode must be installed if there isn't one there already unless the only power to the detached building is a single or multiwire branch circuit. Size the grounding electrode conductor the same way you would size a similar conductor at the main service, using the discussion on this topic in Chapter 8, but based on the size of the feeder conductors instead of the main service conductors.

With the neutral isolation assured, any fault current returns to the service on the equipment grounding conductor. This conductor has a different sizing table in the *NEC* (Table 250.122), which is based on the size of the largest overcurrent protective device ahead of the wire. For 15-amp and 20-amp protective devices, size this conductor at 14 and 12 AWG respectively. For larger circuits up to 60 amps,

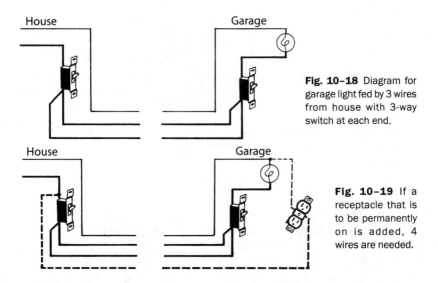

Fig. 10-18 Diagram for garage light fed by 3 wires from house with 3-way switch at each end.

Fig. 10-19 If a receptacle that is to be permanently on is added, 4 wires are needed.

use a 10 AWG wire; for still larger circuits up to 100 amps, use an 8 AWG wire, and for even larger circuits (up to 200 amps) use a 6 AWG wire. These sizes pertain to copper wires; refer to the table in the *NEC* for other circuits or if you are using aluminum grounding wires.

Disconnecting means You must supply a local disconnecting means for all separate buildings and structures fed from another building on the property, even an accessory building to a single-family dwelling. For comparatively simple residential outbuilding applications, the disconnecting rules boil down to the following:

- Ungrounded wires supplying a load intended to stay energized, such as a receptacle, must pass through a disconnecting means located at a readily accessible point nearest the point of entrance.

- A snap switch is a permissible disconnecting means, but a three-way switch with no identifiable OFF position cannot be used. To preserve the functionality of three-way switch control from both locations, place a two-pole snap switch near the point of entrance and electrically ahead of the mating three-way switch placed at its expected location, and interrupt both travelers at once.

- The switches associated with a single source of supply, such as a single branch circuit, must be grouped, although they needn't be as close as adjacent snap switches in a two-gang box.

- Each switch must be marked with its function. If that function is obvious, such as the overhead light, *NEC* 110.22 allows some basis for omitting this marking. However, if you do provide the marking you won't be challenged.

Suppose you install a receptacle that will supply a freezer, and the owner wants to be assured that it won't be turned off inadvertently. Assume there will also be a light controlled from the house and the garage using three-way switches. Mark the two-pole switch set in the travelers ahead of the three-way switch in the garage LIGHT. Run the receptacle feed through a single-pole snap switch in another box near the two-pole switch, or even in the same box, and perhaps at an odd height, say 3 feet above the floor. Over both switches install weatherproof covers as required that preclude inadvertent operation. Finally, mark the second switch RECEP DISC or similar.

Chapter 11
CABLE—NONMETALLIC-SHEATHED AND METAL-CLAD

CABLE CONSISTS OF TWO OR MORE wires assembled together and enclosed in a covering. The two most common types are nonmetallic-sheathed cable and metal-clad cable. For descriptions of their characteristics, including coverings and size nomenclature, see pages 34–36. Approved locations and installation methods for the two types are described in this chapter.

NONMETALLIC-SHEATHED CABLE

Nonmetallic-sheathed cable is light in weight, simple to install without special tools, and costs less than other kinds of cable. It is often called "Romex," which is a trade name of one manufacturer. There are three kinds, which the *NEC* calls Type NM, Type NMC, and Type NMS.

Type NMS is comparatively new. The "S," which stands for signaling (as of the 2008 *NEC*), refers to additional signaling, data, and communications conductors. Per *NEC* 334.116(C), these other conductors must be firmly separated from the power conductors by the cable sheath, and thereby respect the system separation rules in the *NEC* (explained on the first page of Chapter 18).

Where used Type NM cable, shown in Fig. 4–5A and described briefly under that drawing, may be used only in *normally dry indoor locations*. Type NMC was developed for use in damp or corrosive locations indoors or out, but not for exposure to the weather, and not to be buried in the ground. Type NMC may be used instead of Type NM anywhere. Its use is required in damp or corrosive locations such as barns and other farm buildings (further discussed in Chapter 17, "Farm Wiring"), damp basements, and so on. Install it as you would ordinary Type NM.

In some localities it may be difficult to locate Type NMC cable. In that case, use Type UF, which is very similar, doesn't cost much more, and may be used wherever Type NMC may be used. In addition, it may be used in wet locations or buried directly in the ground. Type UF is discussed in more detail in Chapter 17 on page 195.

Building restrictions   There was a dramatic change in the allowable uses of Type NM cable in the 2002 *NEC*: Type NM can be used in any building permitted to be of Types III, IV, or V construction, even if actually constructed as Type I or II, provided the cable is located behind the same sort of thermal barrier normally required for ENT (Fig. 12–7) in highrise buildings. There is no *NEC* waiver for buildings with a sprinkler system, as is the case for ENT; however, building codes generally permit more extensive buildings to be constructed as Type III, IV, or V if a full sprinkler system is in place.

This was first time the *NEC* put building construction types at the center of a rule governing the use of a wiring method. Refer to Annex E in the *NEC*, which extracts this information, for the complete descriptions. They are summarized as follows:

■ *Type I*—All structural members are noncombustible (or limited-combustible) and have fire ratings generally of three or four hours (less in some cases) depending on the specific usage.

■ *Type II*—All structural members are as in Type I but the fire resistance generally drops to two or one hours (less in some cases) depending on the specific usage.

■ *Type III*—All exterior bearing walls are noncombustible (or limited combustible) and have fire ratings of at least two hours, but interior structural elements can be of approved combustible material.

■ *Type IV*—All exterior and interior bearing walls are noncombustible (or limited combustible) and interior columns, beams, girders, arches, trusses, floors, and roofs are of heavy timber construction without concealed spaces.

■ *Type V*—Buildings constructed of approved combustible material that for some structural elements is subject to a minimum one-hour fire resistance rating.

For about thirty years, nonmetallic-sheathed cable could not be used anywhere in any building that exceeded three floors above grade; in contrast to ENT, this restriction applied whether or not the wiring was concealed behind a fire finish. The only exception was in the case of a one- or two-family building, where it could be used regardless of total height. This restriction became the subject of intense and continuing national debate. Some jurisdictions imposed further restrictions, and some others removed the height limitations entirely. Be sure to review your local code because the permitted locations for Type NM cable are often subject to controversy in local code adoption proceedings. This change won't matter very much for single-family homes, but it may influence the choice of wiring method in many highrise condominiums and light commercial applications, in addition to many larger applications beyond the scope of this book.

Ampacity restrictions   Although the individual conductors must, by *NEC* rule, have a 90°C temperature rating, the final allowable ampacity must not exceed that given in the 60°C column (Refer to Table 4–1). This restricts the number of cables that are "bundled" for longer than 24 inches. This includes routing more than one cable through a succession of single bored holes; the NEC requires the use of derating penalties if multiple cables are run together through thermal insulation or if more than two cables with two or more wires pass through a common opening in wood framing that is caulked or fire- or draft-stopped. You need to consider this requirement in a basement, for example, where you might be tempted to run large numbers of cables back to the panelboard through a set of holes lined up through the floor joists and ending at the panel. To make matters worse, a special restriction (which also applies to any wiring method) prevents making use of the general permission to round up to the next higher overcurrent protective device if a branch circuit supplies multiple receptacle outlets. As a practical matter, putting NEC 310.15(B)(3)(a) together with 240.4(B)(1) under these restrictions means that for common applications you can't put more than four cables together through a common set of holes.

Removing insulation   Remove the outer jacket of the cable for 10 to 12 inches. You can use a knife, cutting a slit parallel to the wires and being careful not to damage the insulation of the individual wires. For Type NM, the cable ripper in Fig. 11–1 is very handy and faster than a knife.

Bending   Bending cable sharply may damage the outer cover. The NEC says that all bends must be gradual so that, if continued in the form of a complete circle, the circle would be at least 10 times the diameter of the cable. Cable must always run in continuous lengths from box to box with no splices except inside boxes.

Fastening   Cable must be fastened to the surface over which it runs every 4½ feet, also within 12 inches of every box. Use insulated straps of the general type shown in Fig. 11–2, rather than the staples illustrated in Fig. 11–17, because it is too easy to damage the cable by driving staples too hard. Connectors and clamps should be snug but not over-tightened to the point of conductor damage. Sometimes such damage does not show up until later and is difficult to repair once the cable is concealed in the walls.

Whether exposed or concealed, cable may either follow the side of wood

Fig. 11–1 Using a cable stripper is the fastest way to remove insulation.

Fig. 11–2 Support cable at least once every 4½ feet. Use straps of this general type. ▶

members or run through holes bored in the approximate center of studs, joists, or rafters. Where the edge of a bored hole or the nearest surface of a cable on the side is less than 1¼ inch from the face of the wood member, a ¹⁄₁₆-inch steel plate (or a thinner plate if listed for this purpose) or a bushing must be installed to protect the cable against future penetration by nails, etc. Feed and pull the cable carefully through bored holes to avoid damaging the outer jacket.

Protecting exposed wiring   If the finished wiring is exposed (for example in unfinished basements), see to it that the cable is protected against later mechanical injury. The easiest way is to run it along the side of a stud or joist. If run at angles to such timbers (unless the cable runs through bored holes), a running board must first be installed (See Fig. 11–3). The NEC is not specific about the size of this running board but the so-called "1 by 2" is fine for the purpose. The cable must never be run across free space, and must follow the surface of the building except when mounted on running boards. In unfinished basements, it may be run through bored holes through the center of joists (Cables 8-3 AWG or 6-2 AWG or heavier may be

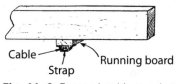
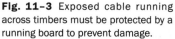

Cable — Running board
Strap

Fig. 11–3 Exposed cable running across timbers must be protected by a running board to prevent damage.

mounted directly across the bottoms of joists without a running board). Where run parallel to framing members, run cable on sides, 1¼ inches or more from face. In accessible attics, cable may run at an angle to joists if protected by guard strips at least as high as the cable, as in Fig. 11–4. Where necessary to provide protection against physical damage, Type NM cable can be drawn into any raceway, with a bushing installed on the raceway ends.

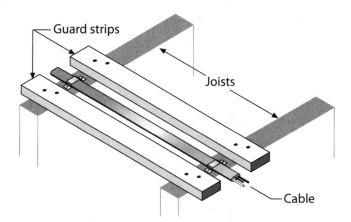

Guard strips

Joists

Cable

Fig. 11–4 In attics, guard strips may be used. If run through bored holes, the cable needs no further protection.

By *NEC* definition, wiring above suspended ceilings with lift-out panels is considered "exposed." In all allowable uses of Type NM cable other than one- and two-family residential, *NEC* restrictions essentially prohibit the use of this cable above suspended ceilings because it generally must not be run "exposed."

Use cable with grounding wire   At one time, Types NM and NMC were made both with and without a bare uninsulated (or green) grounding wire in addition to the two, three, or four insulated circuit wires, but the style without the grounding wire is no longer manufactured. See Figs. 4–5A and 4–5B. In Chapter 10, the diagrams and explanations of wiring procedures are simplified by leaving out the grounding wire. However, for all new installations using nonmetallic-sheathed cable, you will use the kind with the bare uninsulated grounding wire. Connections for the bare grounding wire are explained below.

Connecting the grounding wire   At the starting point in the service equipment, connect the bare grounding wire to the neutral strap, or separate grounding bus, in the cabinet. If the circuit originates in a separate panelboard that is *not part of the service equipment*, the bare grounding wire *must* be connected to a separate equipment grounding bus that is grounded to the cabinet and *not* to the neutral because the neutral is insulated from the cabinet at such locations. At all boxes, connect the insulated wires as if no grounding wire were involved. At each outlet or switch box, the bare wire must be properly connected. If two ends of bare wire enter the box, cut another piece of bare wire a few inches long or use a commercially available green insulated or bare pigtail assembled to a screw. Connect all three ends solidly together—you can use a solderless connector (Fig. 4–18). Ground the opposite end of the short wire to the box itself. You can use a clip (Fig. 11–5), or install an extra screw in one of the unused holes in the box. That screw may be used only for grounding the bare wire.

Receptacles in boxes   If a receptacle is to be installed in the box, run a bare wire (or insulated *green* wire) from the *green* terminal of the receptacle to the junction of the other bare wires. Your solderless connector will then join four ends of bare wire. See Fig. 11–6.

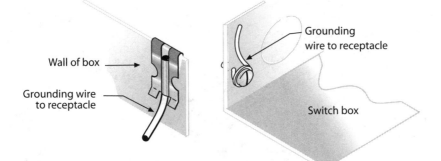

Fig. 11–5 The grounding wire to the green terminal of a receptacle may be grounded to the box using either a grounding screw or grounding clip as shown at left above.

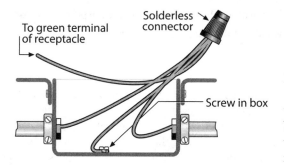

To green terminal of receptacle

Solderless connector

Screw in box

Fig. 11-6 How to install the bare grounding wire of nonmetallic-sheathed cable.

However, many UL listed self-grounding receptacles are available with special mounting screws held captive by wire springs for connecting the receptacle to flush-mounted boxes, providing an effective bond between the receptacle yoke and the metal box. (When using nonmetallic boxes there is no need to connect the bare wire to the box, but all the bare wires must be connected to each other and to the green terminal of the receptacle.) If a receptacle is installed in a surface-mounted box such as a handy box in a basement, so that the mounting strap or yoke is in good solid contact with the box, the wire from the green terminal is not required provided at least one of the fiber retention washers holding the screw to the yoke is removed, assuring grounding continuity. Regardless of circumstances, the bare wires in the cables must always be connected to each other and to the box. Be sure to fold all the bare wires well back into the box so they cannot later touch a live terminal on the receptacle.

Importance of a continuous ground   No matter what kind of boxes are used, the grounding wire must be installed so that it will be continuous from box to box all the way back to the service equipment. Removing a receptacle or other device from the box must not interrupt the continuity of the wire. This is in the interest of good grounding, which means a greater degree of safety. Instructions for conducting a continuity test are given in the next paragraph.

Testing the installation   Even experienced electricians who have confidence in their work test their installations. You should test your work before it is covered up and before receptacles, switches, and fixtures have been installed so that any corrections can be made while it is still accessible. Do your testing *before* the power

Fig. 11-7 A continuity tester. Use only on de-energized circuits. It is also handy for testing fuses.

supplier has energized your installation. There are two tests. The purpose of the first is to check for accidental grounds, and the purpose of the second is to make sure the conductors are continuous.

For both tests all you need is a continuity tester similar to that shown in Fig. 11–7. It is a "penlight" flashlight adapted for testing. Before each use, test the tester by touching the alligator clip on the flexible lead to the projecting prod at the lamp end. If the tester is working, the lamp should light.

Testing for accidental grounds   Conduct this first test after the rough wiring is completed. In new construction that means work at the panelboard is completed and all permanent splices in the branch circuits are completed. Conduct this test *before* the power supplier has energized your installation. At the panelboard, be sure all the circuit breakers are off or the fuses removed. At each switch location, temporarily connect together the hot leg and the switch leg, either by using a solderless connector or just by hooking the bare wires together. Now go to each lighting or receptacle outlet and connect the alligator clip of the tester to the switch leg supplying the fixture (or the hot leg to the receptacle), and connect the prod on the tester to the equipment grounding conductor (there should be one at every outlet) as follows:

▓ If the box is nonmetallic—simply touch the tester prod to the bare or green grounding wire.

▓ If the box is metal—the grounding wire or cable armor should have been connected to the box, so simply touch the tester prod to the box.

In either case, the test light should *not* go on. This establishes that the hot wires throughout the job are "clean"—not accidentally grounded.

Testing for continuity   For this test, use a wire jumper (consisting of short wires that you cut for this purpose) to *temporarily* connect the two "hot" terminals (black, or black and red) and the neutral grounded (white) terminal together at the panelboard. Turn all of the circuit breakers on, or insert all the fuses. Now return to each outlet and use the continuity tester to test between the hot (black) wire and the grounded (white) wire at each outlet. The test lamp should light, showing there is continuity in both sides of the circuit back to the panelboard. When testing is completed, be sure to remove the jumper at the panelboard.

METAL-CLAD CABLE (MC)

Metal-clad cable (Type MC) is available in three forms (interlocking-armor, corrugated, and smooth); however only the first form is generally used for installations covered in this book. It has virtually driven its predecessor, Type AC (armored cable) out of most markets. Unlike Type AC cable, covered later in this chapter, the armor does not generally qualify as an equipment grounding conductor, and therefore we usually see this cable assembly shipped with a green insulated wire sized for equipment grounding.

Although the jacket does have to be grounded at terminations, the separate equipment grounding conductor carries the principal grounding responsibility. Recently some manufacturers have solved the grounding continuity problem across interlocking MC cable armor by using a bare, fully sized aluminum equipment

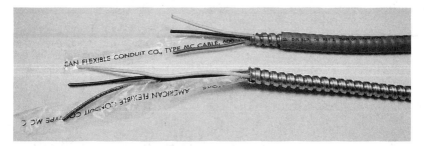

Fig. 11-8 Interlocking-armor Type MC cable. The version at the top has an outer nonmetallic jacket and is listed for direct burial.

grounding conductor against the armor and on the outside of the plastic wrap that goes over the circuit conductors. This product can be used similarly to Type AC cable, without a separate grounding conductor entering the box. Type MC cable can be manufactured with an overall nonmetallic jacket. In this form it may carry a listing indicating it is suitable for direct burial or other uses in wet locations (Fig. 11-8).

Cutting   The safest and quickest method of cutting armored cable is to use a patented cable cutter as shown in Fig. 11-9. It employs a circular saw blade to cut the armor, and has a stop to prevent cutting into the conductors. Make the cut about 10 inches from the end so you will have plenty of wire in the box for connections. If a cable cutter is not available, use a hacksaw, holding it at a right angle to the strip of armor (Fig. 11-10). Be careful to saw only through the armor without touching the insulation of the wires or the bonding strip. This is not easy, so it helps to make some practice cuts on scrap cable. Try bending the cable sharply at the point where it is to be cut. It is difficult to saw completely through a single turn of armor, but usually if the center part of the strip is sawed completely through and the edges partly through, a sharp bend will break the armor and a sharp twist will remove the short end (See Fig. 11-11).

Bends and Supports   Bends in interlocking-armor cables must be no sharper than the point at which the bend, if continued into a circle, would have a diameter 14 times that of the cable diameter. Type MC cable can be fished between access points. Otherwise, support it at least every 6 ft, a more liberal distance than for armored cable. The small branch-circuit sizes, consisting of four or fewer conductors and not larger than 10 AWG, carry an additional requirement for support within 12 in. of a termination. As in the case of cables generally, when you run Type MC cable through holes in structural

Fig. 11-9 A cable cutter offers the safest and quickest method of cutting metal-clad cable. ;

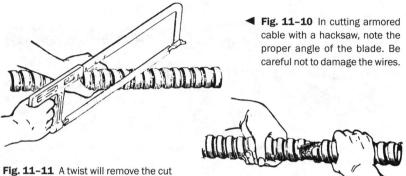

◀ **Fig. 11–10** In cutting armored cable with a hacksaw, note the proper angle of the blade. Be careful not to damage the wires.

Fig. 11–11 A twist will remove the cut end of the armor. Let about 8 inches of wire extend beyond the armor. ▶

framing, whether joists or studs, the *NEC* considers those runs supported. However, at the final support point where the cable enters a box, take care to do more than just cradle the cable in the nearest hole through a framing member. At terminations cables need to be "secured" and not just "supported." That means using a clip or staple that actually prevents the cable from moving in any direction.

Connectors   Type MC cable must terminate in connectors designed for the particular cable involved. Unlike Type AC cable, the *NEC* doesn't require anti-short bushings for Type MC cable, although some leading manufacturers do make them available for that purpose, and many contractors choose to use them anyway. They are not required because the throat designs of listed connectors keep the conductors away from the cut edges of the armor. In addition, since these connectors may have to handle ground-fault currents, they are tested with their designed cable types for this duty. Those tests have no validity beyond the cable types actually tested. Make sure the connector you have in mind was tested for the cable you are installing. Many connectors for Type AC cable are not suitable for Type MC cable.

For metal-clad cable and armored cable, the problems concerning correct colors of wire are exactly the same as with nonmetallic-sheathed cable. As with nonmetallic cable, open splices are not permitted. The cable must be in one piece from box to box. If a splice is necessary, make it in a junction box, as shown in Fig. 10–11.

ARMORED CABLE (AC)

Armored cable (Type AC) is called "BX" by many people, though that is the trademark of one particular manufacturer. It is quite simple in construction (See Fig. 11–12). The wires are thermosetting or thermoplastic, each wrapped in a spiral layer of tough paper. The spiral armor is made of aluminum or galvanized steel and is strong but flexible. Between the paper and the armor there is a bare aluminum bonding strip or wire. This, together with the armor itself, serves the same purpose as the bare grounding wire in nonmetallic-sheathed cable. The armor itself is a very poor grounding conductor compared with copper because the spiral turns of

Bonding strip

Fig. 11–12 Armored cable consists of 2 to 4 wires protected by a layer of tough paper and flexible aluminum or galvanized steel armor. Note the bonding strip under the armor.

the armor do not make good electrical contact with each other. The bonding strip is necessary to assure low resistance and a good ground. Although largely driven off the market by Type MC, it is still used, and you will encounter it extensively in existing structures. It is installed in a similar manner as Type MC cable, but with some important differences covered here.

Installation   Armored cable may be used only in permanently dry locations. Don't use it outdoors. It is cut the same way as Type MC cable. However, the use of anti-short bushings are mandatory for these cables, and these are supplied with the cables (Fig. 11–13).

Fig. 11–13 Always use a fiber or plastic anti-short bushing at the end of the cable between wires and the armor. It provides protection against the sharp edges of the armor that might puncture a wire and cause a ground or a short circuit. They are not required with Type MC cable.

In some cables an overall paper wrapping may make it difficult to insert the anti-short bushing. To make room for it, unwind the paper a few turns under the armor, then give it a sharp yank and it will tear off inside the armor (See Figs. 11–14 and 11–15). Then insert the anti-short bushing as shown in Fig. 11–16. Figure 11–17 shows a cross-section of a piece of cable with the anti-short bushing properly inserted.

Leave an inch or two of the bonding strip projecting beyond the end of the armor. Insert the anti-short bushing as described above, then fold the bonding strip over the bushing and back along the outside of the armor. The bonding strip will help hold the bushing in place until the connector is installed.

Then install a connector as shown in Figs. 9–12 and 9–13 (unless you are using boxes with cable clamps, which make the use of cable connectors unnecessary). The connectors used with armored cable are similar to those used with nonmetallic-sheathed cable except that at the end that goes into the box there are "peepholes" through which the color of the anti-short bushing can be seen. This allows the inspector to see that the bushings have been installed. If you don't use the bushings, your job won't be approved by the inspector.

Installing armored cable   After you have prepared the ends of the armored cable with bushings and connectors, install the cable as described for nonmetallic-sheathed

Fig. 11-14 In removing the paper, first unwind a few turns inside the armor.

Fig. 11-15 A sharp yank tears off the paper inside the armor, and makes room for the anti-short bushing.

Fig. 11-16 Insert the anti-short bushing between wires and armor.

Fig. 11-17 A properly installed anti-short bushing.

cable on pages 127-. Like nonmetallic-sheathed cable, armored cable must be supported every 4½ feet (not 6 ft, as for Type MC) and also within 12 inches of each box. It may be supported using straps, but staples are more frequently used; see Fig. 11-18. The better ones are plated to resist rust. Drive the staples home with a hammer, but not hard enough to damage the cable.

Anchoring for grounded connections Be especially careful in anchoring the connectors to outlet and switch boxes. The connector must be tightly clamped to the armor of the cable. After inserting the connector into the knockout of the box, drive the locknut down tightly enough to bite down into the metal of the box. This makes a good electrical connection so that, in case of accidental grounds, current can flow through the armor from box to box. The white wire in the cable is grounded. The grounding strip is also grounded. If the black wire at some point where the insulation is removed accidentally touches the armor or the box, the result is the same as touching the white wire. It is a ground fault and causes the fuse protecting that circuit to blow or the circuit breaker to open—a signal that something is wrong.

Fig. 11-18 Staple for supporting armored cable.

Grounding terminals of receptacles in boxes The green grounding terminal of the receptacle must be *effectively* grounded. But with armored cable there is no need for the extra grounding wire that is part of nonmetallic-sheathed cable. Instead, the armor of the cable and the bare aluminum grounding strip under the armor serve the same purpose. On the other hand, interlocking-armor Type MC cable, like NM cable, will often have an insulated grounding conductor, and that conductor is handled just as the one for NM cable.

If the mounting yoke of the receptacle is in firm and solid metal-to-metal contact with the box, as in surface wiring, no further action is necessary. In flush work the plaster ears on the yoke of the receptacle usually prevent it from resting directly on the box. As a result the only metallic contact between the box and the receptacle is through the small mounting screws, and that is not good enough. Unless you use the specially designed self-grounding receptacles designed for use without the grounding wire in metal boxes (See page 128), you must install a short length of wire from the green terminal of the receptacle to the box, using either of the methods discussed for nonmetallic-sheathed cable on pages 128–129.

Testing the Installation   Perform the same tests described earlier in this chapter on page 135. In the test for accidental grounds, use the second test (starts with "If the box is metal") which calls for touching the tester prod to the metal box, because only metal boxes are used with armored and metal-clad cable assemblies.

Chapter 12
TUBULAR RACEWAYS

RACEWAYS ARE PROTECTIVE WIRING ENCLOSURES through which conductors are pulled or otherwise inserted after the conduit is installed. The wires and cable may also be removed from conduit if necessary. This chapter covers common tubular raceways, through which wires are pulled from end to end. The *NEC* recognizes many other raceways, including forms into which wires may be installed through removable covers, such as multioutlet assemblies (Fig. 13–23). The most commonly used types of tubular raceways are rigid metal conduit (RMC), rigid nonmetallic conduit made from polyvinyl chloride (PVC), intermediate metal conduit (IMC), electrical metallic tubing (EMT), electrical nonmetallic tubing (ENT), flexible metal conduit (FMC—also referred to as "flex" or by the trade name "Greenfield"), and liquidtight flexible metal conduit (LFMC) and liquidtight flexible nonmetallic conduit (LFNC). This chapter explains general installation procedures, followed by details specific to each of these tubular raceways.

CHOOSING AND INSTALLING TUBULAR RACEWAYS

For raceway sizes there are specific *NEC* requirements for each type, while the general information given below about box depth, bending, and anchoring applies to several types. In general, these raceways may be broadly divided into three groups: nonflexible heavy-wall types, all of which have the term "conduit" in their name; nonflexible thinwall types, all of which have the term "tubing" in their name, and flexible types, all of which have the term "flexible" in their name. All raceway wiring methods must be listed, which means they must bear the label of a recognized testing laboratory. Review Chapter 1 for examples.

Calculating correct raceway size   The size of raceway you will need to use is determined by the raceway type and the dimension and number of wires. Whatever type of tubular raceway is used, *NEC* Chapter 9, Table 1 restricts the wire fill for all such installations to certain percentages of the raceway cross-sectional area. Refer to Tables 4 (for raceway dimensions) and 5 (for wire dimensions) to make the calculations. If all the wires are the same size, *NEC* Annex C has the calculations already done. Note that these tables are specific to each type of wiring method, because the actual cross-sectional area of each wiring method differs. The fill tables in *NEC* Chapter 9 and in Annex C occupy seventy pages, so they cannot be reproduced here.

Box depth with all raceway types   When installing any kind of raceway, be sure to use boxes deep enough so that the knockouts on the sides or ends of the box are sufficiently near the back to allow room for the conduit behind the thickness of the wall or ceiling finish material.

Bends in raceway   No run (as the raceway from box to box is called) may have more than the equivalent of four quarter-bends (a total of 360 degrees) in it. Until you gain experience you may find it best to bend the raceway first, then cut to the required length, because the finished piece after bending may turn out to be a fraction of an inch too long or too short.

Anchoring with pipe straps   Support the raceway using pipe straps. It must be supported within 3 feet of every box, with additional supports not more than 10 feet apart for electrical metallic tubing and for trade sizes 1/2 and 3/4 rigid metal conduit and intermediate metal conduit. You can find support spacing requirements for larger rigid and intermediate metal conduits in Table 344.30(B)(2) of the *NEC*. Nonmetallic raceways and flexible raceways have different support requirements, covered later in the chapter.

Wires in tubular raceways   All wires must be continuous, with no splices permitted inside the run. The raceway is first installed, and the wire is pulled into it later. See page 144 for instructions on pulling wire into raceway.

NONFLEXIBLE HEAVY-WALL CONDUIT
Location requirements and other details of the three types of heavy-wall conduit are described here. Thinwall and flexible methods are covered later in this chapter.

Rigid metal conduit (RMC)   Generally steel, though it may be made of aluminum or other metals, including stainless steel and brass, rigid metal conduit (Fig. 12–1) differs from water pipe in that it is carefully inspected to make sure the inside is entirely smooth to prevent damage to the wires as they are pulled into the pipe. It has a rust-resistant finish inside and outside.

Fig. 12–1 Rigid metal conduit looks like water pipe but differs in several important aspects. It usually comes in 10-foot lengths and each length bears a UL label.

   Where used   Rigid metal conduit may be used either indoors or outdoors, but it may require supplementary protection in corrosive locations such as in cinder fill. Supplementary protection (asphalt paint is one possibility) must be used across the interface when the conduit runs between concrete embedment and soil, such as through a foundation wall below grade.

   Sizes   RMC has approximately the same dimensions as equivalent sizes of standard weight water pipe. Actual dimensions of standard sizes of raceways do not exactly correspond to their common nominal trade sizes. For this reason, and for compatibility with countries using the metric system, the *NEC* no longer refers to raceways as "1-inch" and so on. Starting with the 2002 *NEC*, "trade size designators" or "metric designators" (depending on the measurement system) are used instead. These designators have no units. Standard sizes of rigid metal conduit are trade sizes 1/2, 3/4, 1, 11/4, 11/2, 2, and 21/2, plus larger sizes used mostly in commercial and industrial work. The actual inside diameter in inches is slightly larger than the corresponding trade size designator.

Bending   If conduit is bent sharply, it will collapse. Bends must be gentle and gradual so the internal diameter will not be reduced at the bend. The *NEC* requires bends which, if made with most benders and if continued into the form of a complete circle, would be at least 8 inches in diameter (measured to the centerline of the conduit) for trade size 1/2; 9 inches for trade size 3/4; 111/2 inches for trade size 1; 141/2 inches for trade size 11/4; 161/2 inches for trade size 11/2; 19 inches for trade size 2; and 21 inches for trade size 21/2. To do a good job, use a bender similar to the one shown in Fig. 12-2, available up to 1 inch (11/4-inch for EMT, covered later in the chapter). For larger sizes, use factory-bent elbows. The allowable radii for all sizes are in *NEC* Chapter 9, Table 2. A separate set of dimensions is included for conduit bent with a hickey, which does not offer as much support for the raceway, and which is inched along the conduit instead of making the bend in one shot (motion).

Cutting and threading   The preferred way to cut rigid metal conduit is to use a hacksaw with a blade having 18 teeth to the inch. Cutting leaves a sharp edge or at least burrs at the cut, which can damage wires as they are pulled into the conduit. Use a pipe reamer to remove these dangerous projections. Thread the conduit using dies similar to those used with water pipe, with a taper of 3/4 inch to the foot.

Anchoring with locknut and bushing   Conduit is anchored to a box by means of a locknut and bushing (See Fig. 12-3). The locknut is used on the outside of the box, the bushing on the inside (See Fig. 12-4). The locknut is not flat, but has teeth on one side; the side with the teeth faces the box. The bushing has a rounded surface on the inside diameter over which the wires slide while being pulled into the pipe. To install, first screw the locknut on the pipe as far as it will go, but don't

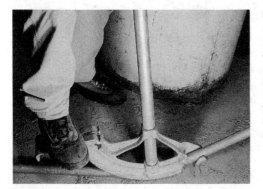

Fig. 12-2 A conduit bender, and method of use. This one bends 1-inch rigid or intermediate metal conduit, or 11/4-inch EMT. For best results, use the heaviest foot pressure you can manage. This bender has a hinged, two-position foot pad allowing heavy, straight-down foot pressure throughout the bending process.

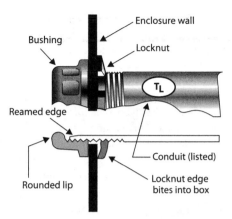

Fig. 12-3 Conduit is fastened to boxes by means of locknuts and bushings, both shown above.

Fig. 12-4 This shows how a locknut and bushing are used. Be sure the locknut is driven solidly home. ▶

do the final tightening yet. Next slide the pipe through the knockout in the box. Install the metal bushing on the inside of the box, screwing it on tightly as far as it will go. Finally, tighten up the locknut on the outside of the box making sure the teeth dig down into the metal of the box. It is common to use nonmetallic bushings, which do an excellent job of protecting the wires but which are not strong enough to anchor the conduit. Use a locknut on both sides of the enclosure wall, and then add the nonmetallic bushing.

Grounding   It is important to tighten the locknut as just described in order to provide a good continuous ground. Many electricians prefer—and some local codes require—the double-locknut procedure just described. After applying the double locknuts you can use either a nonmetallic or metallic bushing. However, if any of the ungrounded wires being pulled in are 4 AWG or larger, the bushing must be nonmetallic, or if metal, it must have a nonmetallic insert in its throat. The *NEC* requires double locknuts (and no concentric or eccentric knockouts) where circuits are over 250 volts to ground. Otherwise, use a grounding bushing as shown in Fig. 8–25 with a wire connected to bond to the main wall of the enclosure. Size this wire just as you would an equipment grounding conductor, as covered under the topic "Grounding" near the end of Chapter 10.

Rigid nonmetallic conduit (PVC)   There are many types of nonmetallic conduit, but the type made from polyvinyl chloride, which is by far the most frequently installed, has its own article in the *NEC*, and it is the only type covered in this book. It is light in weight, easy to install, and has excellent moisture and corrosion resistance. When calculating fill for PVC, be sure to include the equipment grounding conductor, required in essentially every run. See *NEC* Article 352 for further details.

Where used   Except for some hazardous locations and for support of fixtures, PVC can be used wherever rigid metal conduit is permitted, and is preferred for some corrosive locations.

Sizes   PVC conduit in "Schedule 40" has the same dimensions as rigid metal conduit, and in the heavier-wall "Schedule 80" it has the same outside dimensions, but reduced inside dimensions that affect the allowable wire fill.

Bending   Couplings and fittings are solvent-cement "welded" to the outside of the conduit. Factory-made bends and elbows are available, or the conduit can be field bent by first applying heat to soften the material and then using a jig to form the bend, followed by applying a wet cloth to cool it. There are several commercially produced bending heaters available using liquid, air, or convection. In the larger sizes, steps must be taken to maintain the round cross section at bends so it does not flatten. There are also bending springs available that support the inner walls of the conduit, allowing the conduit to be bent using conduit benders at room temperature.

Cutting   Any fine-toothed saw can be used to cut the material, and there are handy blade-type cutters for the smaller sizes. The cut ends should be reamed inside and out.

Intermediate metal conduit (IMC)   Made of steel only, IMC has slightly smaller wall thickness and larger inside diameter than rigid metal conduit.

Where used   In general, IMC is recognized for the same uses as rigid metal conduit.

Cutting and bending   Cutting, threading, and bending are done the same as for rigid metal conduit, except that benders providing side wall support, similar to the EMT bender shown in Fig. 12–2, must be used because the thinner walls must be supported to prevent collapse during bending. Threaded ends will appear a little different, as the tops of the first few threads will be flat, but properly cut threads will mate with standard conduit fittings.

TUBING (THINWALL NONFLEXIBLE RACEWAYS)

Two types in common use, one metallic and one nonmetallic, are covered in this chapter.

Electrical metallic tubing (EMT)   This is also called thinwall conduit, although it is not properly described as conduit. It is shown in Fig. 12–5.

Where used   EMT may be used either indoors or outdoors.

Sizes   In trade sizes 1⁄2 through 11⁄2, it has the same *inside* diameter as rigid metal

Fig. 12–5 EMT cannot be threaded. It is much lighter than rigid conduit. It comes only in 10-foot lengths.

conduit. The *outside* diameter for these sizes is much smaller than for rigid conduit because the wall is much thinner; in fact the wall is so thin that it cannot be threaded. In trade sizes 2 and larger, EMT has the same outside diameter as rigid metal conduit.

Cutting and bending   A hacksaw with 32 teeth to the inch is the most convenient tool for cutting EMT. The tubing must be reamed after cutting. Bend it as you would rigid metal conduit, but it requires more care (and pressure) to keep it tightly against the inner radius of the bender to avoid kinking.

Anchoring with couplings and connectors   Lengths are coupled together and connected to boxes with special pressure fittings—coupling and connector of one type are shown in Fig. 12–6, together with a typical setscrew fitting. Be sure the end of the tubing goes all the way into the connector, against the shoulder stop. Tightening the nut (or setscrew) securely clamps the tubing into the fitting.

Electrical nonmetallic tubing (ENT) As PVC is the usual nonmetallic counterpart to RMC, electrical nonmetallic tubing is the counterpart to EMT. It is

Fig. 12-6 A compression connector and coupling (left and center) used with thinwall conduit. The connector on the right, by comparison, uses a setscrew instead of a compression ring to secure the tubing.

made of the identical PVC in trade sizes up to 2 inches, but in a corrugated wall construction that allows it to be bent by hand without the application of heat. It can be supplied in continuous lengths from a reel. It is even available as a prewired assembly with specified conductor combinations already pulled in place. However, it is not a cable, and it is subject to all the normal restrictions for raceways, including the 360-degree bend rule. It must be supported every 3 feet, and within 3 feet of terminations. It cannot be used outdoors or for direct burial, however, it can be used in cases where it runs in concrete, even if the concrete is below grade.

Since the outside diameter and chemical composition of this product is the same as for PVC rigid nonmetallic conduit, you can use the same solvent-welded fittings. But there are snap-on fittings that are much quicker to apply (Fig. 12-7). For commercial wiring use either PVC or plastic boxes and plaster rings, although metal boxes are acceptable as long as you don't forget to ground them. ENT is also available in a prewired configuration.

You can use the ENT wiring method either exposed or concealed in low-rise construction. However, in buildings that exceed three floors above grade, it must never be exposed, even in the first three floors. Instead, in other than fully sprinklered buildings, it needs to be behind a thermal barrier that has at least a 15-minute finish rating as defined in listings of fire-rated assemblies. In the case of walls, this is fairly easy to arrange because most ½-inch drywall used in commercial construction carries this rating. The same holds true above a drywall ceiling. However, if there is a suspended ceiling (common in commercial occupancies), check with the building inspector. The support grid and the ceiling panels need to be identified as a combination for this duty. For example, having 15-minute panels would do no good if the T-bars dumped those panels onto the floor after 11 minutes of fire exposure.

The first floor is defined as the one with at least half its exterior wall area at or above grade level; one

Fig. 12-7 This shows the corrugations on ENT, which allow it to be bent by hand (a "pliable" raceway according to the NEC definition). A snap-on connector has been added, which will lock in place when it goes into one of the knockouts in the box.

additional floor level at the base is allowed for vehicle parking or storage, provided it is not designed for human habitation. ENT can also be used in highrise buildings (those over three floors above grade) without the use of a thermal barrier if the entire building has a complete fire sprinkler system in full compliance with *NFPA 13, Standard for the Installation of Sprinkler Systems.* The sprinkler system must cover all floors, not just the area where the use of ENT is being considered.

FLEXIBLE RACEWAYS

Three types in common use are covered in this chapter. Two are metal, one being covered with a nonmetallic, waterproof jacket. The third type is also waterproof, but instead of having a nonmetallic jacket over a metal core, it is entirely nonmetallic in cross section.

Flexible metal conduit (FMC)   This wiring method, shown in Fig. 12–8, is available in both steel and lightweight aluminum. It is commonly referred to as "flex" or by the trade name "Greenfield." Flexible metal conduit is similar to the armor of armored cable but the wires are pulled in after the flex is installed.

Where used   This wiring method must never be used in wet locations. FMC is seldom used for a complete wiring system, but it is widely used where some flexibility or movement is required, such as for motors.

Sizes   The smallest for general use is trade size 1/2, but it has about a 3/4-inch outside diameter. Trade size 3/8, which looks like smaller sizes of Type

Fig. 12–8 Flexible conduit is installed in the same way as armored cable, but wires are pulled into place later.

AC or some MC cables, is permitted for connecting single appliances and fixtures. Refer to *NEC* 348.20(A) and Table 348.22 for size and wire fill restrictions for this size product.

Anchoring with connectors   Install flexible metal conduit as you would armored cable. Use connectors of proper size, and then pull the wires into place.

Grounding   You must also install a grounding wire, which may be bare, or insulated if green. But in lengths up to 6 feet (measured to include all flexible wiring methods in the grounding return path), where flexibility is not required and vibration is not present, and where the conductors are protected at 20 amps or less, the flex may serve as the equipment grounding means without an additional grounding wire. Refer to the grounding topic near the end of Chapter 10 for the minimum grounding wire sizes. This wire is connected just as the bare grounding wire is connected in nonmetallic-sheathed cable. Ground the green terminals of receptacles as with nonmetallic-sheathed cable.

Testing the installation   Test as for nonmetallic-sheathed cable and armored cable as described on pages 129–130 in the previous chapter. In the test for accidental grounds, use the second test (starts with "If the box is metal . . ."), which calls for touching the tester prod to the metal box.

Liquidtight flexible metal conduit (LFMC)   This material is similar to ordinary flexible metal conduit, plus it has an outer liquidtight nonmetallic sunlight-resistant jacket. It is commonly called "Sealtite," which is the trade name of one manufacturer. It is covered by *NEC* Article 350. Figure 12–9 shows both the material and the special connector that must be used with it. In order to properly seat the connector, take care to cut the conduit squarely. Part of the connector goes inside the conduit, making a good connection for grounding continuity; part of it goes over the outside, forming a watertight seal. The plastic content in the outer jacket limits the amount of heat it will stand. In general, size your wires so the current they will carry does not exceed the 60°C ampacity column limits (in Table 4–1, Type TW wire) because higher temperatures will soften the jacket. You can exceed those limits if the product is marked accordingly. The product is made and can be used in the smaller ⅜ trade size for applications similar to those where ⅜ trade size flexible metal conduit can be used (and with the same wire fill).

Fig. 12–9 Liquidtight flexible conduit and the special fittings used with it.

In general, as in the case of flexible metal conduit, you need to install a separate equipment grounding conductor when you use this product. Similarly, there is a limited exception for small circuits running with not over 6 feet of flexible wiring in the total grounding return path, and with a similar restriction against use where flexibility is required—however in this case the exception is considerably more complex. The limit is 20 amps in the ½ and ⅜ trade sizes, and 60 amps in the ¾, 1, and 1¼ trade sizes. Larger sizes cannot be used for this purpose regardless of circuit size. There is another complication. For many decades this wiring method was widely—and in many locations exclusively—available as an unlisted product. The unlisted versions of this product do not provide the grounding continuity of the listed products (along with other deficiencies). The listed varieties have a strip of copper wound into the convolutions, and only those varieties qualify for the limited grounding path allowances. Use of the listed product is an *NEC* requirement; make sure that what you are being sold is actually listed.

Liquidtight flexible nonmetallic conduit (LFNC)   This material has the same function as its metallic precursor, but has a completely nonmetallic wall. It is available in three forms, two of which are relatively uncommon and are not covered here. The "B" type, with integral reinforcement within the conduit wall, has become a very popular wiring method. As in the case of the metallic version, you have to be sure the enclosed wires don't run above 60°C, unless the product is marked accordingly.

Both this product and the metallic version can be used outdoors, and even directly buried if listed and marked for this duty. Where exposed to sunlight, it must be marked for this duty as well.

INSTALLING WIRES IN CONDUIT

Assume that you have installed the conduit for outlets A, B, C, and D of Fig. 10–2 in Chapter 10. The installation will look as shown in Fig. 12–10. You are ready to pull the wires into place. For a short run with just two small wires, they can probably be simply pushed in at one end and through to the next outlet. If the runs are longer or have bends, fish tape must be used.

Using fish tape   Fish tape is a highly tempered steel tape about 1/8 inch wide and 1/16 inch thick. As purchased, it will have a loop formed at one end. The tape is flexible enough to go around corners, but stiff enough not to buckle when pushed into conduit. Push the tape into the conduit until the loop appears where the wires are to enter the conduit. Push the bare wires through the loop of the tape, and then twist them back upon themselves as shown in Fig. 12–11. Insulating tape may be used as shown to help hold the wires. If the loop on the end of the tape breaks, you can't form a new one by just bending, because the highly tempered tape often breaks. Heat the end of the tape with a blowtorch, then let it cool slowly in air; this will soften the end so a new loop can be formed.

Wiring at boxes   In pulling the wires into outlets of Fig. 12–10, pull in short pieces from A to B, separate pieces from B to S-B, from B to C, and from C to D. But if the wires from A to B to C, for example, do not require connections at B, you would feed a length in at A, through B, and on to C, leaving a loop in the box at B to facilitate future changes. Follow the rules on this topic on page 103 regarding the length of wire to be left at the box opening. Outlet S-B will contain a switch; use only colored wire because *NEC* 200.7 permits a reidentified white wire (explained on page 112) *only in cable* to be run as the supply to a switch.

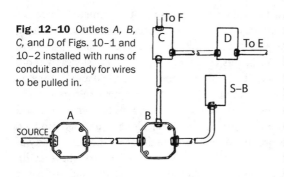

Fig. 12–10 Outlets A, B, C, and D of Figs. 10–1 and 10–2 installed with runs of conduit and ready for wires to be pulled in.

Grounding receptacles   Ground the green terminal to the box unless you are using the self-grounding receptacles mentioned on page 129 that make this unnecessary.

Testing the installation   Test as for armored cable as described at the end of the previous chapter.

Fig. 12–11 Attaching wires to fish tape. Continue the tape over the hook.

Electrical tape

Chapter 13
MODERNIZING OLD WIRING

THE WIRING THAT WAS INSTALLED in a house many years earlier, or even as recently as a decade ago, might not be adequate for the job it is called upon to do today. Some thoughtful analysis will help you decide whether a complete rewiring job is in order, or whether a less expensive approach will serve the purpose.

If the wiring does not include an equipment grounding conductor, either in the form of a separate grounding conductor or a metallic raceway or cable armor, as is often the case for wiring over forty years old, consider completely rewiring all circuits.

Assuming an equipment grounding conductor is present, is the wiring inadequate because you are using too many lights? Too many floor lamps? Too many radios and TVs? That is seldom the case. The wiring usually is inadequate because you have added many electrical appliances that were not considered or perhaps were not even on the market at the time of the original wiring job. The installation does not provide enough circuits to operate a wide assortment of small kitchen appliances, plus range, water heater, clothes dryer, room air conditioners and other heavy appliances. Some of these operate on 240-volt circuits, which may not be available; others operate at 120 volts but, when plugged into existing circuits, they overload those circuits. In addition, the service entrance equipment may be just too small for the load.

To analyze the problem of your particular house, ask yourself this: If you disconnected all the *appliances*, would you have all the *lighting* circuits you need? The answer is probably yes, which means that your rewiring job is simplified. You will still have to rewire the house, but probably not as completely as first appeared necessary. Proceed as if you were starting with a house that had never been wired, but leave the existing lighting circuits intact. (These lighting circuits, of course, will include many receptacles used for small loads like a vacuum cleaner, radio, and TV, but not the receptacles for kitchen or laundry appliances.) This chapter discusses the overall plan for modernizing an existing installation, including upgrading to a larger service and adding new circuits. It also offers problem-solving approaches to the specific challenges you are likely to encounter. Everything in previous chapters describes the wiring of buildings while they are being built, which is called "new work." This chapter describes "old work," which is the wiring of buildings after they have been completed.

PROBLEMS OF "OLD WORK"

There is little difference between old and new work, except that in old work there are a great many problems of carpentry. The problem is to cut an opening where a fixture is to be installed, and another where a switch is to be installed, and then to get the cable inside the wall from one opening to the other with the least amount of work and without tearing up the walls or ceilings more than necessary.

One house to be rewired may be five years old; another, a hundred years old. Different builders use different methods of carpentry. Every job will be unique. No book can possibly describe all the methods used and all the problems you will meet. Watch buildings while they are being built to get an idea of construction at various points. In old work, good common sense is of more value than many pages of instruction.

In general, old work requires more material because it is often wise to use 10 extra feet of cable to avoid cutting extra openings in the walls or to avoid cutting timbers. Many problems can be solved without cutting any openings except the ones to be used for outlet boxes and switch boxes. Others require temporary openings in the wall that must later be repaired. Techniques are given in this chapter for running cable behind walls and ceilings, and for installing boxes, switches, receptacles, and outlets in both lath-and-plaster and drywall construction.

Caution: Even in an abandoned, vacant, or isolated building, always test for voltage before starting any work. It is possible, with other buildings on the property, or with a two-meter service for yours and another occupancy, that there might be feedback which would be surprising at the least, or hazardous at worst.

WIRING METHODS

The wiring method you will use in old work will be either nonmetallic-sheathed cable with grounding wire, or metal-clad cable (or even armored cable)—whichever is the custom in your locality. Nonflexible raceway methods cannot be used unless the building is being practically rebuilt. In a few areas, flexible raceways are used. In this case be sure to install the grounding wire along with the circuit wires as discussed on page 139.

Knob-and-tube wiring is permitted by *National Electrical Code* (*NEC*) Article 394 only for extensions to existing work. The materials (knobs, tubes, and loom) are difficult to find and the building wires currently manufactured do not fit the knobs. If you need to extend a knob-and-tube circuit, or pick up a portion with a new house run, extend the wires encased in loom (a flexible woven fibrous nonmetallic tubing) continuous from the last knob into a new accessible junction box, and change to the cable wiring method you are using. Consider, however, that it is far better to completely rewire these circuits.

Concealed knob-and-tube, as a wiring method, has no equipment grounding conductor carried with it. Over the generations, *NEC* provisions have changed to the point that it is almost impossible legally to wire anything without grounding it. Until the 1993 *NEC*, you could go to a local bonded water pipe to pick up an equipment grounding connection, and then extend from there with modern wiring methods. *NEC* 250.130(C), which governs this work, requires that the equipment grounding connection be made on the equipment grounding terminal bar of the supply panelboard, or directly to the grounding electrode system or grounding electrode conductor. You are unlikely to be searching for a method of grounding concealed knob-and-tube wiring in a steel-frame building. Rather you will be attempting this in old wood-frame buildings, probably residential. In such occupancies, even if the water supply lateral is metallic, the water piping system ceases to be considered as an electrode beyond 5 feet from the point of entry. This means fishing into the basement. If you can fish a ground wire down into the basement, you can fish a modern circuit up in the reverse direction and avoid the entire problem.

It's true that some geographical areas have more extensive use of slab-on-grade construction, and here interior water piping is sometimes permitted to qualify as electrodes because the pipes extend to grade for the minimum threshold distance of 10 feet, and thereby allow interior connections. But in almost every case, trying to extend knob-and-tube wiring is like trying to erect a modern structure on a rotten foundation. The same issues apply to antique Type NM cable with no equipment ground, and to old Type AC cable with no bonding strip to short the convolutions.

Add to these problems the fact that beginning with the 1987 *NEC*, concealed knob-and-tube wiring cannot be used in wall or ceiling cavities that have "loose, filled, or foamed-in-place insulating material that envelops the conductors." This effectively means that such cavities cannot be insulated, because you'd have to open all the walls to install board insulation products, and if you'd do that, you'd have no reason to consider trying to save this wiring method.

The writer considered including specific instructions at this point in how to perform an extension of concealed knob-and-tube wiring—this would include instructions on positioning knobs and cleats, end fittings for the new wiring method, loom, soldering, etc.—and decided against it. If you find concealed knob-and-tube wiring, rewire it. If instead you choose to extend it, try to find some old loom from elsewhere on the job, and slip it over the individual conductors to the last knob. Loom is flexible nonmetallic tubing just big enough to slide over an individual conductor. Consider yourself lucky if the piece you find is still flexible. Cut it just long enough to enter a box at the nearest feasible point, and bring it in, one wire per trade size ½ knockout (using a Type NM cable connector) or per cable knockout. Be sure the wire enters the box at least 6 inches beyond the end of the loom. If you have a steel box, snip the web between the two knockouts per *NEC* 300.20(B) so you don't create inductive heating around what is probably antique and fragile Type R insulation. Even better, use a nonmetallic box and cover for this purpose.

INSTALLING NEW SERVICE AND CIRCUITS

Install a new service of at least 100-amp capacity, with a new outdoor meter, beginning with the insulators on the outside of your house and ending with new service equipment inside the house. Let the equipment contain, as a minimum, branch circuit breakers to protect all the new circuits that you are going to install, plus a few spares. Be sure to consult your power supplier about how you plan to proceed. Review Chapter 8 for calculating service size.

When you have finished installing your new service and circuits, you will have no power on your new circuits but will still have power on the old circuits. *Call your power supplier to disconnect all power on the outside of your house.* Get along for a day or two without electric power while you reconnect the old circuits into your new equipment.

When all the work has been completed, have your power supplier install the meter and connect the power to your new service if they are willing to do so. In some jurisdictions, particularly on upgrades of smaller residential services, the electrician on site must do the cut and reconnection. Because the service drop will be live, this is potentially dangerous work for an unqualified person to perform. Some power suppliers will require a licensed electrical contractor to install this phase of the work. Before you commit yourself to doing this work, be sure to review local practice with your power supplier and with your local inspectional authority. When this final connection is completed, you can then enjoy the advantages and pleasures of a newly rewired home.

Installing new circuits   Using the old-work techniques described later in this chapter, install two small-appliance circuits, or better, one 3-wire circuit. Install the laundry circuit and the bathroom receptacle circuit. Install individual circuits for heavy appliances such as range, water heater, clothes dryer, and furnace motor. You will be connecting each of these circuits to the breakers in the new equipment.

Provide arc-fault circuit interrupter (AFCI) protection if feasible. Review the AFCI discussion at the end of Chapter 5, page 50148, and note that the *NEC* now requires all wiring "modified, replaced, or extended" in the covered areas to have AFCI protection.

Installing the service equipment   Whether you retain or replace the old service equipment panelboard, you must disconnect the incoming service wires from the present equipment. Follow Option A below if you will continue to use your old panelboard. Follow Option B if you plan to discard your old panelboard entirely. With either option, you must disconnect the ground wire from the old equipment and remove it. You must install a new ground wire from the new equipment to the ground, and an equipment grounding conductor between the new and old equipment enclosures. In the discussion that follows, circuit breakers are mentioned but fused equipment may be used.

Option A—Using the old panelboard   Your new equipment will contain a main breaker, breakers for all the new branch circuits you have installed, plus a few spares for future circuits, and one 2-pole, 30-amp or larger breaker from which you will run wires to the old equipment. Let's assume your present installation is three-wire, 120/240-volt.

Disconnect the incoming service wires from your old equipment. If they are in conduit, remove the conduit. If they are in cable, remove it so it doesn't enter the cabinet at all. Then run three-wire cable from the old equipment to the new. The two hot wires of the cable run from the 30-amp or larger breaker mentioned in the previous paragraph to the terminals in the old equipment to which the old service wires were connected. The white wire in the cable runs from the grounded busbar in your new equipment to the old equipment. The white wires in your old equipment will present a problem. If the old equipment was installed comparatively recently, the grounded busbar in it may be bonded to the cabinet either (1) by a special bonding screw—remove it and the busbar will then be insulated from the cabinet; or (2) by a flexible metal strap bonded to the cabinet with the other end connected to the grounding busbar—disconnect it from the grounding busbar, which will then be insulated from the cabinet (it will be best to cut off the bonding strap completely). Then connect the white wire from the grounded busbar in the new equipment to the now-insulated busbar in the old equipment.

If your old equipment was installed many years ago before the schemes described in (1) and (2) above were in use, the grounded busbar was probably bonded directly to the cabinet with no way of insulating it. Remove it if possible, but in any case remove the white wires from it and install them in a new, insulated, grounded busbar which you must purchase and install. Make sure it has enough terminals of the right size to accommodate all the white wires.

If your old equipment is two-wire, 120-volt, it is functionally obsolete and should be discarded; Use Option B, below, in this event.

This completes the wiring. Your old equipment is now connected to the new equipment. If your old equipment contained *main* breakers or fuses, leave them as they are; they are not required but will do no harm. But if there are fuse clips that

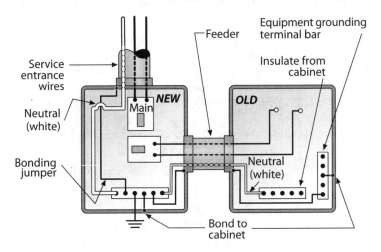

Fig. 13-1 Here the original equipment is not a part of the service equipment. It is only a "load center" for the original circuits.

seem deteriorated or terminals that appear to be in poor condition, remove them.

If the original equipment served only lighting circuits (including receptacles), the wires from the old to the new equipment might get by as 10 AWG protected by a 30-amp breaker. But if the old equipment served and continues to serve 240-volt loads as well, install 6 AWG wires protected by a 60-amp breaker. See Fig. 13–1.

Option B—Replacing the old panelboard   Remove the original service wires. Disconnect all branch circuit wires, but don't cut off any of the wires. If the trim or cover of your present equipment has a hinged door in it, or if any openings in it can be closed off, remove and discard the interior (breakers and bus, or fuseholders, terminals, neutral bar, etc.), install knockout seals in any unused openings, and use the enclosure as a junction box. If this is impractical, proceed as follows: if the wiring used armored or nonmetallic-sheathed cable, remove the locknut inside the cabinet, pull the cable connector out of its knockout, and temporarily screw the locknut on its connector. If the wiring was in conduit, remove the bushing inside the cabinet, pull the conduit out of the knockout, and place the bushing on the end of the conduit. After doing this on each branch circuit, remove the old equipment completely.

The new equipment will contain a main breaker, plus other breakers to protect each of the branch circuits, old and new, plus a few spares. But the wires of the old circuits will not reach the new equipment. Where the old equipment used to be located, install a junction box (an empty steel cabinet of convenient size, 8 × 8 inches or larger as needed, with steel cover). Run the wires of the old circuits into this junction box, using the original cables with their connectors, or the original conduits.

Then run separate cables from the new equipment to the junction box, one for each circuit and of the same size as the original wires. Connect the new wires to the old, black to black, white to white, using "wire nuts" or other solderless connectors. See Fig. 13–2. The junction box will not contain a grounding busbar (although one could be added), and the white wires in the box must be carefully

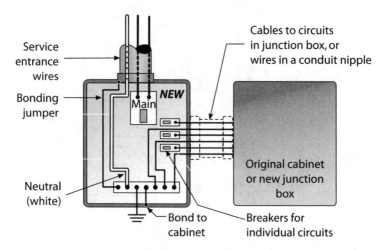

Fig. 13–2 Here the cabinet of the old equipment, if retained, serves only as a junction box.

insulated from each other and totally insulated from the box, but the box itself must be grounded.

Although circuit modifications generally invoke compliance with all current AFCI requirements that would apply to new wiring in equivalent circumstances, the *NEC* waives this rule in cases like this. The only conditions are that no additional outlets or devices may be installed, and that the extensions made to existing conductors must not exceed 6 ft.

INSTALLING BOXES AND RUNNING CABLE

The *NEC* requires that boxes enclosing flush devices must be at least $^{15}/_{16}$ inch deep. Lighting outlet boxes may be as little as ½ inch deep as shown in Fig. 13–3, but deeper boxes should be used wherever possible. Cable is simply pulled into the walls and anchored to the outlet and switch boxes. Each piece must be a continuous length from box to box. In all wiring the *NEC* limits any gaps between boxes and the surrounding wall material to ⅛ inch or less. If it is not practical to cut openings close to the exact size of the box, use patching plaster to fill gaps after the wiring is finished.

Fig. 13–3 This ceiling pan, with a volume of 6 cubic inches and no required fill adjustments for an internal cable clamp, can just accommodate a single 14-2 AWG Type NM cable, or a 14, 12, or 10-2 AWG Type AC cable.

Selecting box location   In locating outlets, consider that all wires must be fished through walls and ceilings. Sometimes by moving an outlet or switch a foot or so, a difficult job of boring through joists or other timbers can be avoided. In old work, the switch and outlet boxes are supported directly or indirectly by either the lath under the plaster or by the drywall, so choose the locations for the openings carefully. A location fairly close to joists and studs is best because there the wall materials are better supported.

Mounting switch boxes in lath-and-plaster construction   In sawing openings for switch boxes in walls and ceilings of lath-and-plaster construction, remember that the length of a switch box is approximately the same as two widths of lath plus the space between the laths. If you remove two complete widths of lath, the mounting brackets on the switch boxes will barely reach the edges of the next two laths, and the laths will split when you drive the screws (1-inch No. 4 flat-head screws are commonly used; the best choice is flat-head sheet metal screws that have the entire shaft threaded). Remove one width of lath completely, and part of another on each end of the opening. Figure 13–4 shows the wrong and the right ways.

Make a mark on the wall approximately where the switch or receptacle is to be located. Bore a small hole through the mark. Insert a stiff wire and probe to make

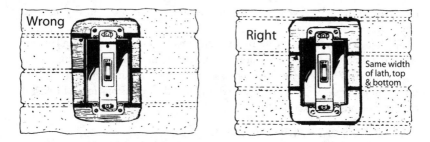

Fig. 13-4 In cutting an opening for a switch box, cut away one whole lath and part of another lath on each side of the one completely cut. This provides a rigid mounting for the box.

sure there is no obstruction and that there is sufficient space all around. Enlarge the hole to locate the center of one lath—this will locate the vertical midpoint of your opening. Then mark the area of your opening, about 2 inches by 3¾ inches. Bore ½-inch holes at opposite corners, and at the center of top and bottom. See Fig. 13–5. The centers of the holes must be on the lines of the outline so that not more than half of each hole will be outside of the rectangle. Unless you watch this carefully, parts of the holes may later not be covered by your switch or receptacle plate.

The holes at the corners are for inserting a hacksaw blade or fine-toothed keyhole saw for cutting the opening. The holes at top and bottom provide clearance for the screws used to mount switches or receptacles. If using a hacksaw blade, wrap one end with tape for a handle. Insert the blade so the sawing is done as you pull the blade out of the wall. Support the plaster with a piece of wood as you pull the blade toward you. If you saw as you push the blade into the wall, you will probably loosen the laths from the plaster, leading to a very flimsy mounting of the box.

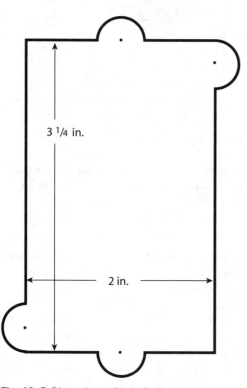

Fig. 13-5 Dimensions of hole for installing a switch box. To save time, make a template. Photocopy this page, glue it to a piece of stiff cardboard, and cut to size.

Mounting switch boxes in drywall construction   Drywall is not sturdy enough to accept screws as is lath-and-plaster construction. Mount the box near a stud (use a stud finder) for better support. Use the template in Fig. 13–5 to cut the opening. Adjust the brackets on the ends of the box so that when installed the box will be flush with the wall surface.

Connecting the cable   In both types of construction, boxes with beveled corners and cable clamps are convenient but seldom practical. The beveled corners usually reduce the capacity of the box below *NEC* minimums, and are seldom used. Bring the cable into the box and tighten the clamps, letting about 10 inches of cable extend out of the box. If using boxes without clamps, install the connector on the cable, let the connector project into the box through a knockout near the back of the box, and install the locknut after the box is installed. Be sure the box is deep enough and the knockout far enough back for the cable connector to clear the inner surface of the wall.

Anchoring switch boxes   The following three methods can be used in both lath-and-plaster and drywall construction:

- Use a box that has special clamps on its outside walls (See Fig. 13–6). After installing the cable, push the box into the opening, then tighten the screws on the external clamps. This makes the clamps collapse, anchoring the box in the wall.

- Use an ordinary box plus the U-shaped clamp of Fig. 13–7. Install the U-clamp with the screw holding it in place unscrewed about as far as it will go. When you slip the box into its opening, the ends of the clamp will expand outward. Tighten the screw holding the clamp to anchor the box ears firmly against the wall.

- Use an ordinary box plus a pair of special straps shown in Fig. 13–8. Insert one strap on each side of the wall opening and push the box into the opening, taking care not to lose one of the straps inside the wall. Then bend the short ends of each strap

Fig. 13–6 This box has exterior collapsible clamps on each side. Push box into opening, tighten screws on sides, and the box will be anchored in the wall. *(Hubbell Electrical Products)*

down into the inside of the box. Be sure they are bent sharply over the edge of the box and lie tightly against the inside walls of the box so they cannot touch the terminals of a switch or receptacle installed in the box, which would lead to grounds or short circuits.

Mounting outlet boxes in ceilings   If there is open space above the ceiling on which the box is to be installed, and if there is no floor above (or there is a floor in

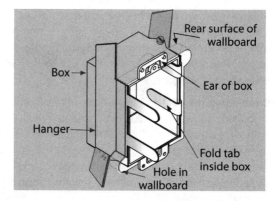

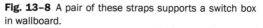

Fig. 13-7 Another way of supporting switch boxes.

Fig. 13-8 A pair of these straps supports a switch box in wallboard.

which a board can easily be lifted), proceed as in new work using a hanger and the usual 1½-inch-deep box. See Figs. 9–8, 9–9, 9–10. The only difference is that you will be working from above.

If all the work must be done from below, the method depends on the ceiling construction, the location of the outlet, and the weight of the fixture. Boxes supporting fixtures weighing more than a few pounds must be fastened to the building structure. One method is shown in Fig. 13–9. For lightweight fixtures the ceiling itself can support the box. Use one of these methods: (1) ½-inch-deep box, surface mounted, supported by a ceiling joist using screws; (2) ½-inch-deep box, surface mounted, supported by fixture stud on bar hanger (Fig. 9–8) which has been poked up through a hole (which the box covers) and laid across the wood laths (Fig. 13–10); (3) ½-inch-deep box, surface mounted, supported by toggle bolts through ceiling, either lath and plaster or drywall (Fig. 13–11); (4) 1½-inch-deep box with ears, flush mounted, supported by a U-clamp (Fig. 13–12). When using a surface-mounted ½-inch-deep box, be sure to select a fixture having a canopy which will cover the box, as shown in Fig. 13–9. In all these cases, fish the cable to the outlet location and fasten it to the box before securing the box.

Remember that if the ceiling is combustible (if it can burn), you must cover the space between the edge of the box and the edge of the fixture with a noncombustible material. Surface raceway manufacturers (Fig. 13–23 being one form of this wiring method) have round boxes for making extensions from ceiling outlets; the flat bases of these boxes can be used for this purpose.

Lifting floor boards   Often a board in the upstairs floor must be lifted to get at the ceiling space. This is no problem if the flooring is rough, as in ordinary attics. But if the lumber is tongued and grooved, care must be taken to avoid marring the floor. The first step is to cut the tongue off the boards. A putty knife cut off short, so the blade is only about an inch long, makes an excellent chisel for the purpose.

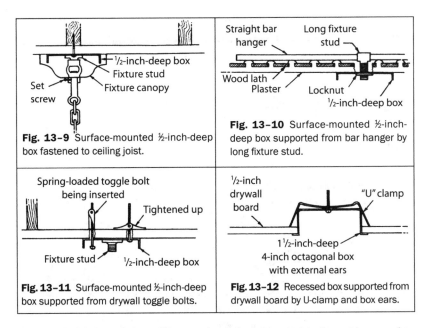

Fig. 13-9 Surface-mounted ½-inch-deep box fastened to ceiling joist.

Fig. 13-10 Surface-mounted ½-inch-deep box supported from bar hanger by long fixture stud.

Fig. 13-11 Surface-mounted ½-inch-deep box supported from drywall toggle bolts.

Fig. 13-12 Recessed box supported from drywall board by U-clamp and box ears.

Sharpen the blade and you will have a chisel about 1 or 1 1/2 inches wide, very thin, but short and stubby, which makes it strong. With this you can get down into the crack between two boards and chisel off the tongue as far as necessary. Then bore two holes in the board as close as possible to joists—see Fig. 13-13. With a keyhole saw, cut across as close to the joist as you can. It is best to cut at an angle so that the board, when replaced, forms a wedge. The board should be removed over the space of at least three joists so when replaced it rests directly on at least one joist. When replacing the board, first nail a cleat to the joist at each point where you sawed across. These cleats must be very solidly nailed so that when the board is replaced there will not be any springiness. If you have a steady hand, you can pocket cut the board with an electric saber saw, and avoid the bored holes that would need to be plugged. In fact, if you are very careful, you can even cut the flooring across the middle of the floor joist, avoiding the need for cleats.

Before taking this step, be absolutely sure you've got the right location. This trick works if you have any line of sight into the joist cavity, even with a mirror. Usually hardwood flooring is installed over a subfloor, which allows the ends of the boards to end randomly; the installer doesn't have to cut them to end halfway across a joist. In this case, make your best guess where you want to open the floor. At the end of one of the existing boards nearest your projected opening, drill a tiny hole (about 1/16 inch) straight down at one corner. Because the hole occurs at two intersecting lines in the floor, it will never be seen. Push a bright colored rod straight down into the joist cavity (a white coat hanger is a good choice). Note from below where it came through, and adjust your measurements if necessary.

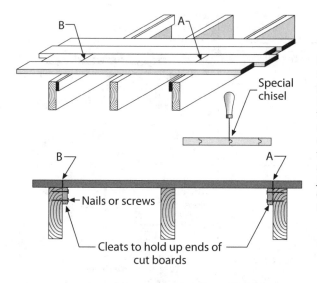

Special chisel

Nails or screws

Cleats to hold up ends of cut boards

Fig. 13-13 Sometimes floor boards must be lifted. If the boards are tongued and grooved, cut off the tongue on each side with a very thin chisel, then saw across next to the joists. To simplify the drawing, the subfloor is not shown.

Temporary openings in walls   Sometimes a temporary opening must be made in a wall so that cable can be fished around a corner. On papered walls, use a razor blade or utility knife to cut through the wall covering to form a top-hinged flap as shown in Fig. 13–14. Before lifting the section of paper, soak it with a wet rag to soften the paste. Thumbtacks will hold the lifted portion out of the way while you make a temporary opening in the drywall or plaster. When the wiring is finished, paste the paper back into place.

Fish tape   See Fig. 12–11 and related text for information about using fish tapes, which are very helpful in old work for fishing cables.

Running cable to two openings on same wall   In Fig. 13–15 cable must run from opening *A* to opening *B*, both in the same wall. Depending on the structure of the building, the cable may run in one of three ways.

Under floor boards   Running the cable as in "Route 1" in Fig. 13–15 is the simplest way. Use this route if you can easily lift the floor boards in the floor above so the cable can be dropped down from above to the locations of opening *A* and opening *B*.

Through the basement   If it is very difficult to get into the ceiling space from above, it may be possible to run the cable down through the basement as shown by "Route 2" in the same figure. If the wall is an outside wall, there will probably be an obstruction where the floor joins the wall. In most cases it is possible to bore upward through this at an angle from the basement. Then push two pieces of fish tape upward through the bored holes until the ends emerge at *A* and *B*. Then by pulling at *A* and *B*, fish the opposite ends of a piece of cable upward until the ends come out at *A* and *B*. If the wall is an inside wall, there may be no partitions in the basement immediately below this wall so that it should be possible to bore straight

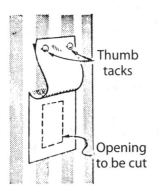

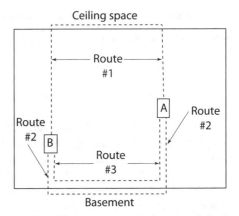

Fig. 13-14 In old work, lift the wallpaper over the area of a temporary opening, as shown here. The top of the paper serves as a hinge.

Fig. 13-15 In running cable from *A* to *B* (both on same wall) there is usually a choice of three routes for the cable. The best route depends on the structure of the building.

upward. Then fish the cable upward to opening *A* and opening *B*.

Behind the baseboard   In the problem above, if it is impossible to run the cable through either the ceiling space above or the basement below, use "Route 3" by running the cable behind the baseboard along the bottom of the wall. First remove the baseboard. Then make a small opening into the wall behind the baseboard directly under *A* and another directly under *B*. Between these two openings cut a channel in the wall into which the cable can be laid. Figure 13–16 shows the completed installation. Wherever the cable crosses a stud, cover it with a 1/16-inch steel plate (or thinner, if listed for this duty) to protect it from future penetration by nails. Finally, replace the baseboard.

From ceiling opening to wall opening   The problem in this case is to run cable from an opening in the ceiling, labeled *C* in Fig. 13–17, around the corner

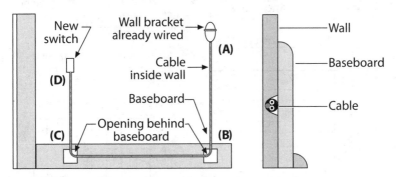

Fig. 13-16 Running cable in a trough cut in the wall behind the baseboard.

at *D*, and down through the wall to opening *E*. It may be a very simple problem or a difficult one, depending on the construction of the house. If the ceiling joists run in the direction shown in the small inset of the same drawing, the problem is greatly simplified. If the floor above is easily lifted, it is then a simple matter to pull the cable in at opening *C*, drop it down at *D* until it comes out at *E*. Even if there is an obstruction at *D*, as is usually the case, it is easy to bore a hole down from above after the board has been lifted.

If it is impossible to lift the floor above, the cable must be gotten around the corner at *D* some other way. An opening must be made into the wall. Sometimes it is made at the point marked *No. 1* by cutting through the wall, and next chiseling away part of the obstruction. Push a length of fish tape into this opening until the end shows up at *C*. Pull it out of *C* until the opposite end is at opening *D*. Then carefully push it down inside the wall until the end shows up at *E*. You then have a continuous fish tape from *C* around *D* to *E*. Attach the cable to the fish tape at one end, pull at the opposite end, and fish it through the wall until you have a continuous cable from *C* to *E*. Sometimes it is easier to do this from the opposite side of the wall, as at point marked *No. 2*, boring upward through the obstruction as shown by the dotted arrow. Then using fish tape, pull in the cable as before. After the cable has been fished, patch the wall and the job is finished.

If the opening *E* is not directly below point *D*, but is to the right or left, run the cable over to the proper point above *E* (if the floor board above can be removed) and drop down. If the flooring cannot be removed, drop from *D* down to the baseboard; behind the baseboard run over to a point below opening *E*, and then run upward to *E*.

If the joists of the ceiling run in the wrong direction, as shown in Fig. 13–18, there is again a choice of routes. If the floor above can be removed, follow *Route 1*, boring holes through the joists through which the cable is to run. If the floor cannot be lifted, make an opening at point *X*, drop the cable down at *X* to the baseboard below, run it behind the baseboard, around the corner to a point below *E*, and from there upward to *E*.

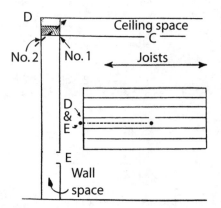

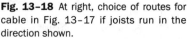

Fig. 13–17 At left, problem in running cable from opening *C* in ceiling, around corner at *D*, to *E* in side wall.

Fig. 13–18 At right, choice of routes for cable in Fig. 13–17 if joists run in the direction shown.

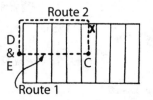

No two houses are alike, so you will simply have to use your common sense in getting around obstructions. Temporary openings often have to be made, and in all cases you will probably use more material than you would for new work. The cost of a few extra feet of cable for a longer route is insignificant compared to the additional time it would take you to follow the shortest route.

INSTALLING SWITCHES, RECEPTACLES, AND OUTLETS

Replace old receptacles if they are damaged or if you wish to change the color. Options for replacing two-wire receptacles are given in this section. Much convenience can be gained from adding new switches to existing outlets and adding new outlets to existing circuits. New outlets should be only for greater accessibility—be sure they are not used to add to the total load on a circuit.

Replacing old two-wire receptacles   Where there is no grounding means in the box, there are three options for replacing two-wire receptacles: (1) two-wire receptacles, (2) three-wire receptacles grounded to the grounding electrode system or the grounding electrode conductor, or (3) GFCI-protected three-wire receptacles. If you use this method, the *NEC* requires that the new receptacles supplied from a GFCI receptacle or circuit breaker must be marked GFCI PROTECTED and NO EQUIPMENT GROUND.

Adding switches to existing outlets   The connections in the present outlet will look a great deal like the left-hand part of Fig. 13–19. There may be more wires in the box than shown, but there will be only two wires connected to the fixture, one white and one black. In the right-hand part of Fig. 13–19 is shown the same outlet after the addition of the switch. To make the proper connections at the fixture, open the black wire splice to the fixture, thus producing two new ends of wire. These two new ends are connected to the two wires in the cable which runs to the switch. The black wire from the fixture is connected to the black wire in the new piece of cable; the black wire of the cable, which runs up to the original outlet box, is connected to the white wire reidentified as black in the new piece of cable. This is contrary

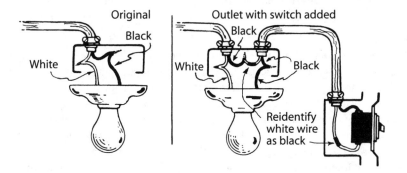

Fig. 13–19 Proper method of adding a single-pole switch to an existing outlet.

to general practice, but is the one case where the *NEC* permits a black wire to be attached to a white. See Fig. 10–4 and the surrounding discussion for an explanation of this *NEC* requirement.

If the switch loop cannot be easily rewired, a neutral must be extended to the switch location, whether single-pole or three-way, as covered in Fig. 10-6 and on page 63.

Adding outlets   Figure 13–20 illustrates an added receptacle outlet, and Fig. 13–21 illustrates an added lighting outlet. The end of the cable—marked TO SOURCE—must be run to an existing outlet box, which contains, in addition to the white wire, a black wire that is *always* hot. If you are in doubt as to whether one of the black wires in any box is always hot, it is a simple matter to check. Turn off the main switch. Remove the cover and the receptacle or fixture from the box from which you plan to run. Take the tape off the connections and leave the exposed ends of wire sticking out of the box. Using two solderless connectors, temporarily connect each wire to one of the leads of a neon tester of the type shown in Fig. 19–2. Now turn the main switch back on. If the tester continues to light regardless of whether the switch controlling the outlet is on or off, you have found your source for the new outlet. *Be sure to turn the power off again before working on the wires.*

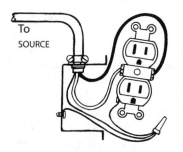

Fig. 13–20 Wiring of a new baseboard receptacle.

Fig. 13–21 Wiring for a ceiling outlet.

Alternatives to conventional outlets   If you don't wish to undertake an extensive modernizing effort, consider the following solutions to several common problems. Note the special requirements on page 117 will apply to added receptacles.

Two-to-three-wire adapters   What if you live in an older house with only non-grounding two-wire receptacles, and you want to use an appliance with a three-wire cord and a three-prong plug? An alternative to replacing or adding outlets is to use a "two-to-three-wire" adapter shown in Fig. 13–22. Note that it has a green terminal lug on its side. Remove the faceplate and then test to establish that the box is grounded. Using the test light shown in Fig. 19–2, insert one wire in the narrow slot in the receptacle and touch the edge of the metal box with the other wire. Then insert the tester in both slots, testing line to line. If the box is grounded the lamp will light both times with equal brilliance. Replace the faceplate, plug in the adapter, and reinstall the screw through the green lug, which is then in contact with the mounting yoke of the receptacle. This procedure is by no means the complete equivalent of having grounding receptacles in your home, but it does permit you to use appliances with three-prong plugs, and does provide some degree of protection *as long as the receptacle yoke is grounded to the grounded metal box.* But if your home is wired

Fig. 13-22 This adapter permits a 3-prong grounding plug to be used with an ordinary receptacle.

using a wiring method that includes an equipment grounding conductor (either metal raceways, or cable armor, or nonmetallic methods with a separate grounding conductor), it would be far better to replace the two-wire receptacles with the grounding types, of course adding a grounding wire from the green terminal of the receptacle to the grounded metal switch box as discussed on pages 128–129.

Never use a two-to-three-wire adapter, or a conventional three-wire grounding receptacle, on a circuit that does not provide an equipment ground. You will be falsely advertising the presence of a grounding connection that does not exist. Only GFCI-protected devices are allowed for this purpose, because GFCI protection provides shock protection whether or not a grounding connection is present at a protected outlet.

Fig. 13-23 Multioutlet assembly provides closely spaced multiple outlets.

Multioutlet assemblies   Few homes have all the receptacle outlets that the occupants would like. The multioutlet assembly in Fig. 13–23 makes outlets available at intervals of 6 to 24 inches as desired. It consists of a metal channel with wires and receptacles already installed. In living rooms the channel can be installed directly above the baseboard with molding added above the channel, so the whole assembly appears to be part of the baseboard. In a kitchen it is installed above the counter at a convenient height for appliances. Similar surface raceways are available in nonmetallic form. They are easily installed, light in weight, and come with a complete line of fittings, switches, and receptacles.

Handy boxes   If exposed surface wiring is acceptable, you can use the "handy" utility boxes shown in Fig. 9–5, using any appropriate wiring method.

Extension rings   Where the new wiring may be permanently exposed, as in basements, it is sometimes convenient to use an extension ring, which is like an outlet box without a back. Remove the fixture of the existing outlet, mount the extension

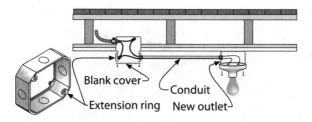

Blank cover
Conduit
Extension ring   New outlet

Fig. 13-24 Using exten-sion rings makes it easy to add to existing out-lets in locations where the new wiring may be permanently exposed.

ring to the flush box, then run cable or conduit for the new run from the extension ring. Replace the fixture on top of the ring. See Fig. 13-24.

Chapter 14
APPLIANCES

WIRING REQUIREMENTS FOR SOME COMMON appliances are explained in this chapter. Appliances are connected to the power supply either by a cord and plug or by permanent wiring. Cord-and-plug-connected appliances include those that are readily portable, such as toasters and vacuum cleaners. Some large appliances are permanently connected, and others come equipped with a cord and plug to facilitate servicing. Kitchen ranges, clothes dryers, and similar appliances are located on individual circuits. Some appliances are fixed—that is, fastened in place through plumbing connections or other installation conditions; examples are water heaters, oil burner motors, and central air conditioners.

When you are in the market for a new major electrical appliance, consider the more energy-efficient models being offered. The initial cost may be higher than for a conventional appliance, but the savings in power used over the life of the appliance will more than make up for the higher price and will help to conserve energy. To encourage the use of more efficient appliances, some power suppliers have incentive programs such as a cash payment for junking an old refrigerator, or a rebate for an approved appliance purchase.

RECEPTACLES

Receptacles are rated in amperes and volts. The volt rating indicates the following restrictions:

- rated at 125 volts—may be used at any voltage up to but not over 125
- rated at 250 volts—may be used only at voltages over 125 but not over 250
- rated at 125/250 volts—may be used only for appliances that operate at 120/240 volts and that require a neutral wire running to the appliance

Figure 14–1 shows a variety of receptacles labeled *A* through *L*. Except for *A*, all 125-volt and 250-volt receptacles have a third opening for the third prong on a three-prong plug for connection to the equipment grounding wire as discussed on pages 72–73.

The 125/250-volt receptacles *I* and *J* have a third opening for the third prong on

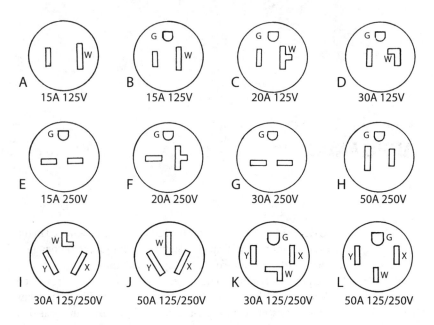

Fig. 14-1 An assortment of receptacles. Those rated at 15 and 20 amps have a diameter of 1.327 inches. Those rated at 30 and 50 amps have a diameter of 2.12 inches (except K and L, which, with 4-wire connections, are even larger in some designs, at 2.38 inches). In the illustrations, the opening marked G is for the equipment grounding wire; the one marked W is for the white circuit wire; those marked X, Y or Z are for the hot circuit wires.

the plug for connection to the grounded neutral wire of the circuit.

Note that a plug that fits A will also fit B or C; a plug made for B will also fit C; but a plug specially made for C will not fit A or B. All of the other configurations are non-interchangeable.

At A is shown what used to be the typical 15-amp receptacle (usually made in the duplex configuration) and is used only for replacements. The receptacle at B is the same as A, except it has the third opening for the third prong on a three-prong plug—for connection of the equipment grounding wire. At C is a similar 20-amp receptacle. At D is a 30-amp receptacle for larger 120-volt loads.

At E, F, G, and H are shown respectively 15-amp, 20-amp, 30-amp, and 50-amp receptacles for loads operating at 240 volts. Note that each is provided with an opening for the equipment grounding prong on a three-prong plug. *Caution:* That prong and that opening must never be used for a neutral wire if the appliance operates at 120/240 volts rather than at 240 volts.

At I and J are shown 30-amp and 50-amp receptacles for appliances operating at 120/240 volts. The 30-amp receptacle is used mostly for clothes dryers, and the 50-amp receptacle for ranges on pre-1996 *NEC* circuits. At K and L are shown three-pole, four-wire, 125/250-volt receptacles for newer dryer and range circuits.

These are not the only receptacles available. There are dozens of others in two-wire, three-wire, four-wire, and even five-wire types. Besides the ordinary variety, there are others designed so the plug cannot be removed without first twisting the plug to unlock it. Some special-purpose types not used in ordinary wiring are designed so that only a plug and receptacle of the same brand will fit each other.

Receptacles come in a variety of mounting methods to fit various boxes and plates, both flush and surface types. The 50-amp receptacle is shown in both the surface-mounting type and flush-mounting type in Fig. 14–2, which also shows a typical plug (on left) with "pigtail" cord attached. These are used mostly in the wiring of electric ranges. A similar 30-amp receptacle is used for clothes dryers.

Fig. 14–2 Ranges are connected to 50-amp, 125/250-volt receptacles of the type shown here. The pigtail is connected to the range, then plugged into the receptacle.

INDIVIDUAL CIRCUITS FOR APPLIANCES

The *National Electrical Code* (*NEC*) rules are quite complicated regarding when an appliance requires an individual branch circuit serving no other load. In general, you will be following the *NEC* rules if you provide a separate circuit for each of the following:

- Range (or separate oven or counter-mounted cooking units)
- Water heater
- Clothes dryer
- Clothes washer
- Waste disposer
- Dishwasher
- An appliance that would otherwise take more than 50 percent of a 15- or 20-amp branch circuit that supplies lighting and/or cord-and-plug-connected loads not fastened in place
- Any 120-volt permanently connected appliance rated at 12 amps (1440 watts) or more, including motors
- Any 240-volt permanently connected appliance
- Any automatically started motor such as a well pump
- Any central heating equipment such as an oil burner or gas furnace (although auxiliary equipment such as humidifiers and zone valves can use the same circuit)

GROUNDING OF APPLIANCES

The following appliances must always be grounded per *NEC* 250.114(3): refrigerator, freezer, air conditioner, clothes washer, clothes dryer, dishwasher, waste disposer, sump pumps, aquariums, personal computers, and fax machines. This grounding is especially important if the appliance is installed where a person can touch both the appliance and the ground or a grounded object. For this reason the *NEC* specifically includes motor-operated tools and electric lawn mowers and hedge clippers, etc. (A concrete floor, even if tiled, is considered the same as the actual earth.) For safety, always ground your appliances.

The one general exception to all of this is listed equipment protected with a system of double insulation. This equipment will be distinctively marked, and its attachment plugs will not have a grounding pole. That is, they will mirror the *A* configuration of Fig. 14–1.

The grounding of ranges and dryers is discussed in specific sections later in this chapter. For other appliances, if they are supplied with a cord that includes a grounding wire, and a plug with a grounding blade fitting a properly installed grounding receptacle, that is all that is required.

If there is no cord and plug, but the circuit wires run directly to the appliance; and if the wiring is armored cable or in conduit, check to make sure the frame of the appliance is grounded to the junction box on the appliance to which conduit or armor is anchored. But if the wiring method is by nonmetallic-sheathed cable, you must use cable with the bare grounding wire, and the bare grounding wire must be connected to the frame of the appliance.

In the case of a water heater on a farm, if the wiring is grounded to a driven ground rod but there is some buried water pipe—no matter how short—you *must* interconnect such pipe with the ground rod. This is an essential step to prevent a difference of voltage between them and to minimize danger from lightning. This is discussed in more detail in Chapter 17, "Farm Wiring."

DISCONNECTING MEANS AND OVERCURRENT PROTECTION

Every appliance must be provided with some means of disconnecting it completely from the circuit and must be provided with overcurrent protection.

Portable appliances   The plug-and-receptacle arrangement is all that is required. The plug and receptacle must have a rating in amperes and volts at least as great as that of the appliance.

Small permanently connected appliances   If the appliance is rated at 300 watts or less (1/8 hp or less), the branch circuit overcurrent protection is sufficient. No special disconnecting means is required. Range hood fans and bathroom exhaust fans are examples of small permanently connected appliances.

Large appliances, not motor-driven   Wall-mounted electric heaters, ceiling-mounted heat lamps, and water heaters are examples in this category. The overcurrent protection must generally not exceed 20 amps, or 150 percent of the rated current if higher than 13.3 amps (next higher standard size being permitted, however). The disconnecting means must be in sight of the appliance (refer to the discussion on page 184 for full coverage of this topic) or be capable of being locked open, such as the type shown in Fig. 16-3. A thermostat having a marked "OFF" position is permitted if within sight and arranged to open all ungrounded conductors.

Large motor-driven appliances   For large motor-operated appliances, the disconnect must be within sight of the appliance, which can be a cord and plug interfacing with a receptacle if it is accessible. A range plug accessible by the removal of a drawer in the base of the appliance is considered accessible.

Appliances with unit switches having marked "OFF" positions are acceptable as is provided the disconnecting function is backed up elsewhere. In a single-family house, this can be the service disconnect, and for a two-family it can be the feeder (or service) disconnect for the occupancy in question. Multifamily (three or more units) occupancies can use circuit protective devices if they are within the unit or at least on the same floor. Other occupancies can rely on the branch circuit protective device, provided it is readily accessible while servicing the appliance.

WIRING OF COMMON HEAVY APPLIANCES

The *NEC* does not restrict the methods used for wiring heavy appliances. Use conduit or cable as you choose.

If the appliance is to be connected by cord and plug, run your cable up to the receptacle, which may be either flush-mounted or surface-mounted. The *NEC* demands that the receptacle be located within 6 feet of the intended location of the appliance; with good planning it should be possible to locate it even closer thus simplifying installation of the appliance.

White wire reidentified for use as black wire   White wire may be used only for the grounded wire. But that wire does *not* run to any appliance operating at 240 volts. Therefore the wires running to a 240-volt load may be any color except white or green. When you use a two-wire cable to connect a 240-volt load, the cable contains one black wire and one white wire, but the white wire must not be used. What can be done? Follow *NEC* 200.7(C) instructions to reidentify the wire as described on page 112.

Provision for grounding of appliances on pre-1996 *NEC* circuits   *NEC* 250.140 makes an important provision that applies only to older circuits. While many appliances must be grounded, in the case of ranges (including counter units and separate ovens) and dryers, their frames may be grounded to the neutral circuit conductor, provided it is 10 AWG or heavier. See *NEC* 250.140. Moreover, for these appliances and no others, you may use service entrance cable with a bare neutral, provided it runs from the appliance directly to the service equipment.

Ranges and dryers are 120/240-volt appliances, and the neutral wire carries current in normal operation. Three-wire service-entrance cable with a bare neutral may be used in wiring 240-volt appliances such as water heaters, etc., to which the neutral does not run, provided the bare wire of the cable is used *only* as a grounding wire.

Ranges   In some ranges, and in all older ranges, a surface burner operates at either 120 or 240 volts, depending on whether it is turned to low, medium, or high heat. The individual burners are connected within the range in a manner that makes it impossible for the neutral wire to carry as many amperes as the two hot wires. For that reason, the wires to the range usually include a neutral that is one size smaller than the hot wires. For most ranges, two 6 AWG plus an 8 AWG neutral are used; for smaller ranges, two 8 AWG with a 10 AWG neutral are occasionally used. Surface burners on most modern ranges operate at 240 volts and have stepless control. The neutral is still needed for operating the clock, timer, oven light, etc.

Run your circuit up to the range receptacle of Fig. 14–2; this is rated at 50 amps, 125/250 volts. The range is connected to the receptacle using a pigtail cord shown in the same illustration. This also serves as the disconnecting means.

Grounding the range   The *NEC* requires that the frame of the range be grounded by means of a separate green (or bare) grounding wire, using a four-wire cord and plug (unless it is permanently connected). On older circuits the range is permitted to be grounded through the neutral wire, in which case a bonding strap is connected between the neutral wire and the frame of the range, and the cord and plug are three-wire. The requirement to use four-wire supplies for ranges took effect with the 1996 *NEC*. Check that the bonding strap is not connected in new installations where a four-wire cord is used.

Receptacle outlet for gas range   Install a receptacle outlet for a gas range. This receptacle, which can be on the small appliance circuit, is for the supply of a gas ignition system, lights, clock, and timer.

Sectional ranges   The trend is away from complete self-contained ranges, consisting of oven plus burners, toward individual units. The oven is a separate unit, installed in or on the wall. Groups of burners in a single section are installed in or on the kitchen counter where convenient. This makes for a very flexible arrangement and permits you to use imagination in laying out a custom-designed kitchen. The *NEC* calls such separate ovens "wall mounted ovens," and the burners "counter mounted cooking units." Here they will be referred to merely as ovens and cooking units or counter units.

Unlike self-contained ranges, ovens and cooking units are considered fastened in place. They may all be either permanently connected or cord-and-plug-connected.

Two basic methods are used in the wiring of ovens and cooking units. Supplying a separate circuit for the oven and another for the cooking unit is one method. The alternate method is to install one 50-amp circuit for the oven and cooking unit combined. Any type of wiring method may be used. Regardless of the wiring method used, the frame of the oven or cooking unit must be grounded.

Where a separate circuit is installed for the oven, use wire with the ampacity required by the load. The oven will probably be rated about 4,500 watts, which at 240 volts is equivalent to about 19 amps, so 12 AWG wire would be suitable. At the oven, the circuit wires may run directly to the oven, but some prefer to install a pigtail cord and a receptacle. Note that because the plug and receptacle are concealed behind an appliance that is fastened in place, the plug will not serve as the disconnecting means as it does when installing a self-contained range.

To wire the cooking units, proceed exactly as for the oven, using no smaller than 10 AWG wire. This size wire has an ampacity of 30 amps and will provide a maximum of 7,200 watts, which will take care of most cooking units. Use a pigtail cord and a receptacle if you wish to make it easy to service the unit.

If you install a single circuit for oven and cooking units combined, it must be a three-wire, 50-amp circuit. Any wiring method may be used, including service-entrance cable with a bare equipment grounding conductor. The receptacles must be the 50-amp type, and may be flush receptacles installed in outlet boxes, or the surface type shown in Fig. 14–2. The circuit will be as in Fig. 14–3. The wires to the receptacles must be the same size as the circuit wires. But the wires from the receptacles to the oven or cooking unit may be smaller, per *NEC* 210.19(A)(3) Exception 1, provided they are heavy enough for the load, not smaller than 12 AWG (10 AWG if used for grounding), and not longer than necessary to service the appliance. The receptacles are not required but they may be convenient for installation. The oven or cooking unit may be connected directly to the circuit wires in the junction boxes. The *NEC* exception permits smaller wires between the junction box and the appliance under the same conditions specified for when receptacles are used.

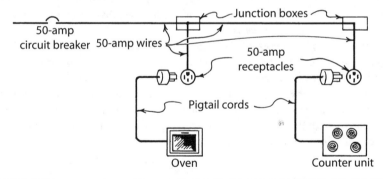

Fig. 14–3 It is common to provide separate circuits for oven and counter units, but both may be placed on one circuit as shown.

Clothes dryers   An electric dryer is a 120/240-volt appliance. Wire as for a range. The *NEC* requires a three-pole, four-wire, grounding-type, 30-amp receptacle (See *K* of Fig. 14–1), and a four-wire pigtail cord similar to that used for a range but with smaller wires. The plug and receptacle serve as the disconnecting means.

Grounding the dryer   For pre-1996 *NEC* circuits only, *NEC* 250.140 permits the frame of the dryer to be grounded to the neutral of the three wires if it is not smaller than 10 AWG. Service-entrance cable with a bare neutral was permitted where it runs from the service location directly to the dryer receptacle. For these older applications only, a three-wire receptacle may be used (configuration *I* of Fig. 14–1).

Clothes washers   Washers are equipped with a cord and plug for easier servicing, so provide a 20-amp grounding receptacle on a 20-amp circuit. No separate switch is required.

Water heaters   *NEC* 210.19(A)(1) requires that loads expected to be on continuously for three hours or more must not exceed 80 percent of the branch circuit rating. Dwellings and farms seldom have such continuous loads, but a domestic water heater is required to be considered in this category by *NEC* 422.13. This means that a 4,500-watt water heater must be on a 25-amp circuit, not a 20-amp. Dividing 4,500 watts by 240 volts equals 18.7 amps, but 18.7 is 93.5 percent of 20. The next higher standard overcurrent device rating is 25 amps, and 18.7 is 75 percent of 25. Therefore, the 25-amp circuit is the minimum. The maximum is given by the 150 percent rule covered earlier; 18.7 × 1.5 = 28 amps; since the next higher standard size is permitted, the more common 30-amp protective device is customarily applied. This additional 25 percent of the continuous loads must be added to feeder and service calculations also, per *NEC* 215.2(A)(1) and 230.42(A)(1).

Look at the water heater terminal box for a marked temperature rating for the branch circuit wires. If there is no marking, the circuit wiring can be TW (or other 60°C wire), but if marked 75°C, then use a wire with an H in its designation (or HH if marked 90°C). For further information, see *NEC* Article 310. Where higher temperature wires are required, it is common to splice the ordinary branch circuit wires in a junction box near the water heater to short lengths of the higher temperature wire extending to the heater.

In some localities, power for heating water is sold at a reduced rate, with the heater connected to the circuit through a special electrically operated switch furnished by the power supplier. The switch connects the heater to the power line only during off-peak hours. Each day for several periods of several hours each, water cannot be heated. If your installation is of this type, do the wiring as already described, except that the wires should start from the power supplier's time switch instead of from your service equipment.

Chapter 15
FINISHING YOUR ELECTRICAL INSTALLATION

ALL THE WIRING DESCRIBED IN previous chapters is done as the building progresses. Installation of switches, receptacles, faceplates, and fixtures—considered finishing work—usually occurs only after the walls have been papered or painted. The finishing portion of the work is a very small and usually simple part of the total job.

INSTALLING DEVICES IN BOXES

Figure 15-1 shows how to mount a switch and its cover plate. The switch is mounted in the switch box using machine screws that come with the switch. The plate is then mounted on the switch using screws that come with the plate. Inspect any switch or receptacle, and you will see that the holes for the screws that hold the device to the box are elongated rather than round. This allows the device to be mounted on a true vertical even if the box was installed crookedly (See Fig. 15-2).

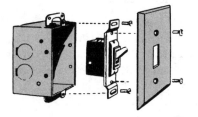

Fig. 15-1 The switch or other device is first mounted in the box, and the plate is then fastened to the device.

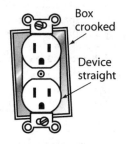

Fig. 15-2 Elongated holes in strap allow devices to be mounted straight in a crooked box.

Adjusting boxes   Cut wall openings accurately to minimize the patching needed after installing the boxes. The perimeter gap between box and wall surface can be no more than 1/8 inch wide. Boxes must be installed so the front edges are no more than 1/4 inch behind the surface of the finished wall or ceiling if these consist of noncombustible materials (materials that will not burn). In all other cases, boxes must be flush with the wall or ceiling surface. Take care, for example, when installing combustible paneling directly over existing drywall—every device box in the room will likely end up in violation. To solve this problem, there are flush box inserts that extend the reach of recessed boxes. These inserts, if made of metal, end up very close to the energized terminal screws on the sides of devices, and putting electrical tape over the device screws is a sensible precaution.

In general, the *NEC* requires flush devices to be held securely at the finished surface [i.e. *NEC* 406.5(A)]. For this reason switches and similar devices have "plaster ears" (See Fig. 15-3) that lie on top of the wall surface and bring the device to the proper level even if the front edge of the box is behind the surface. You can easily break off the ears if they are in the way. If the wall surface is too damaged to support the plaster ears, insert enough No. 6 washers under the yoke to support it even with the wall.

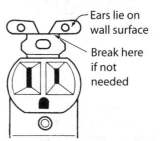

Fig. 15-3 Plaster ears are helpful in aligning a device flush with the wall surface.

Preventing loose connections   Here is a tip that can save you a lot of trouble: after connecting the wires to a switch or receptacle, fold the wires around to tightly hug the back of the device. Then when you push the device into place, the resistance offered by the stiffness of the wires is taken by the back of the device and not by the wiring terminals. Many loose connections, especially at receptacles, result when an acceptably tightened terminal screw is loosened as the device is pushed into place in the box.

Mounting faceplates   When mounting faceplates, don't draw up too tightly on the mounting screws. Common plastic plates are easily damaged. Plates for duplex receptacles have only a very narrow strip of material between the two openings; overtightening will crack this bridge and ruin the plate. This is another reason to mount flush devices so they are "held rigidly at the finished surface."

HANGING FIXTURES

It is impossible to describe all the possible ways to hang light fixtures. It all depends on the construction of the particular fixture. With the general information given here you should have no trouble because fittings to suit the particular fixture are usually supplied by the manufacturer.

Connecting the wires   On some fixtures, one wire is white, the other black. Often both are the same color, but one has a colored tracer thread woven into the covering of the wire. The white wire, or the wire with the tracer, always goes to the white wire in the box. The other wire goes either to the black wire in the box or to the switch. Connect the wires from the fixture to the wires in the box using solderless connectors.

Mounting on boxes   The *NEC* requires that all fixtures be mounted on outlet boxes. Very simple small fixtures can be mounted directly on the boxes using screws supplied with the fixture (See Fig. 15-4). Somewhat larger fixtures often use a special strap supplied with the fixture. This method is shown in Figs. 15-5 and 15-6. Surface-mounted fluorescent fixtures may have a cable or conduit enter the end of the fixture, and suspended fixtures may be cord-and-plug-connected to a receptacle in a ceiling box.

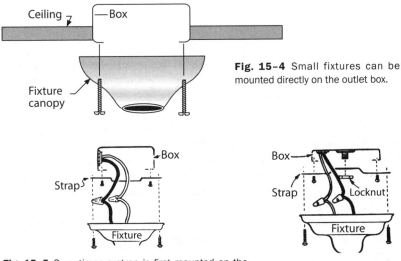

Fig. 15-4 Small fixtures can be mounted directly on the outlet box.

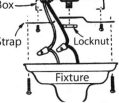

Fig. 15-5 Sometimes a strap is first mounted on the box, then the fixture is mounted on the strap. Straps with grounding terminals are available for nonmetallic boxes.

Fig. 15-6 Another method of mounting a fixture.

In the event you mount a large fluorescent fixture directly over a recessed box, with the fixture supported by the structure of the building and not the box, the *NEC* requires access to the wiring in the box without removing the fixture. This may mean punching a large (3-inch or so) hole in the back of the fixture lined up with the box opening to provide this access. Don't forget to install a bonding jumper from the box to the fixture to maintain grounding continuity.

Still larger fixtures are commonly hung directly on a fixture stud mounted in the back of the outlet box, or on the fixture stud that is part of the hanger on which the box is supported. The "stem" of the fixture is threaded to fit the fixture stud. Slide the canopy down the stem. When the work at the outlet box is finished and all connections made, slide the canopy up to conceal the wiring. See Fig. 15-7.

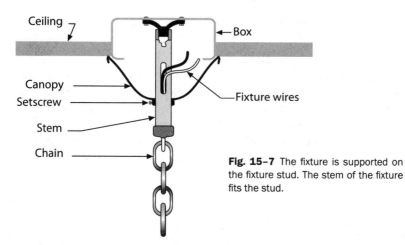

Fig. 15-7 The fixture is supported on the fixture stud. The stem of the fixture fits the stud.

Wall brackets   For mounting a wall bracket fixture not exceeding 6 pounds in weight, two No. 6 machine screws may be used. Use a switch box as if a switch or receptacle were being installed. With the wall bracket you will find a mounting strap, a short threaded nipple, and a knob (or a long heavy screw in place of the nipple and knob). Mount the strap on the box, then install the fixture on the strap. Sometimes a fixture stud is mounted in the back of the box in place of the strap. If the wall is combustible (if it can burn, as in the case of wood), the space from the box to the edges of the fixture per *NEC* 410.13 must be covered with noncombustible material such as sheet metal. Surface raceway manufacturers have round boxes for making extensions from ceiling outlets; the flat bases of these boxes are conveniently punched and drilled for outlet box mounting and can be used for this purpose. Other manufacturers have ceiling medallions that may fit acceptably.

Recessed fixtures   A portion of a recessed fixture must be installed at the rough-in stage of the work, rather than after the ceiling finish is on and painted. Most residential recessed fixtures have a connection box spaced away from the fixture to allow connection to ordinary branch-circuit wiring. However, some are marked with a temperature rating for the branch-circuit wiring which may require that you use a wire with an H or HH in its type designation (TH, THW: 75°C; THHN, XHHW: 90°C). Be sure the fixture is spaced at least 1/2 inch from the joists or other combustible material (except at points of support) and that it will not be blanketed by thermal insulating materials, which will prevent the escape of heat (unless the fixture is marked "Type I.C.," meaning it will not overheat even when blanketed with thermal insulation). Do not install lamps larger than the size marked on the recessed fixture because the additional heat will damage wire insulation and could start a fire.

Track lighting   Decorative fixtures to "wash" walls with light, highlight art objects, or illuminate wall-hung paintings can be installed anywhere along a surface-mounted or flush track. Many types of fixtures are available for a great variety of decorative lighting functions. Power to the track is supplied from a ceiling outlet.

Porch fixtures   A fixture installed on the ceiling of an outdoor porch must be marked as suitable for a damp location. It is installed the same as an indoor fixture. If a wall bracket fixture is exposed to the weather, be sure it is listed, marked as suitable for a wet location, and installed so that moisture cannot penetrate behind the fixture and enter the building walls.

PART 4 SPECIAL WIRING SITUATIONS AND PROJECTS

Chapter 16
ELECTRIC MOTORS

FOR SAFETY, EFFICIENCY, AND ECONOMY, the size and type of motor you choose should be based on the requirements of the job it will be performing. This chapter discusses the work capacity, power consumption, and installation of the types of motors commonly used in homes and on farms. Special attention is given to safety factors, including grounding and protecting with overload devices.

WORK CAPACITY OF MOTORS
The capacity of a motor to do work is based on its horsepower, starting and overload capacity, and speed. Choose a motor size that is appropriate to the task.

Horsepower   A motor is rated in horsepower (hp). One horsepower is defined as the work required to lift 33,000 pounds one foot (33,000 foot-pounds) in one minute. One horsepower is equal to 746 watts. Typical horsepower ratings are included in the descriptions of motor types in this chapter.

Starting capacity   Motors can deliver far more power while starting than they can at full speed. The proportion varies with the type of motor; some types have starting torques four or five times greater than at full speed. Naturally the amperes consumed during the starting period are much higher than while running at full speed. In selecting a motor, it is important to consider the start-up load. The motor will soon overheat if too heavy a starting load prevents it from reaching full speed.

Overload capacity   Almost any good motor will develop from 1 1/2 to 2 times its normal horsepower for short periods after coming to full speed. Thus a 1-hp motor is usually able to deliver 1 1/2 hp for perhaps 15 minutes, 2 hp for a minute, and usually even 3 hp for a few seconds. No motor should be deliberately overloaded continuously, for obvious reasons. But this ability of the motor to deliver more than its rated horsepower is very convenient. For example, 1/2 hp may be just right for sawing lumber, but when a tough knot is fed to the saw blade the motor can instantly deliver 1 1/2 hp and then drop back to its normal 1/2 hp after the knot has been sawed.

Replacing gasoline engines with electric motors   Unlike an electric motor, a gasoline engine has no overload capacity. That is why it is often possible to replace a 5-hp gasoline engine with a 3-hp electric motor. If the gasoline engine always runs smoothly, and if it seldom labors and slows down, it can be replaced by an electric motor of a lower horsepower. But if the engine is always laboring at its maximum power, the motor it replaces it should be of the same horsepower as the engine because no motor will last long if it must run overloaded.

Speed of electric motors   The theoretical speed of a motor depends on the number of poles and the line frequency. The most common speed for a 60-Hz motor is theoretically 1,800 rpm. Actually the motor runs at a little over 1,750 rpm while idling and somewhere between 1,725 and 1,750 rpm while delivering its rated horsepower. When overloaded the speed drops still more. If overloaded too much, the motor finally stalls. The speed of ordinary ac motors cannot be regulated by rheostats or switches, but there are variable frequency drives for this purpose that control the natural speed of the motor. Special variable-speed motors are obtainable, but they are expensive special-purpose motors and are not described here.

How temperature rise affects operation   A 10-hp motor made today isn't much bigger than a 3-hp motor made fifty years ago. This is possible because of advances made in the heat-resisting properties of insulations on the wires used to wind the motor, and insulations used to separate the windings from the steel in the motor. Other advances have led to reduced air gaps and to improved magnetic properties of the laminated steel used in the pole pieces. Today's motors will run much hotter without being damaged, but motors should always be installed where they will have plenty of air for cooling.

Motors have stamped on their nameplates a "service factor" ranging from 1.00 to 1.35. Multiply the horsepower by the service factor. The answer tells you what horsepower the motor can safely deliver continuously in a location where the temperature is not over 40°C (104°F) while the motor is not running. That means the motor might develop a temperature of over 100°C (212°F), the boiling point of water, but it will not be harmed.

Most motors of 1 hp or larger have a service factor of 1.15; smaller motors have a higher service factor, some as high as 1.35. A motor works most efficiently and lasts longer if operated at its rated horsepower.

POWER CONSUMED BY MOTORS

A motor is a rotating machine that converts electrical energy to mechanical energy. A motor delivering 1 hp—746 watts—is actually consuming about 1,000 volt-amperes from the power line. Some of the difference is heat loss in the motor due to friction in the bearings and other factors. The motor also has inherent magnetizing current over and above the portion of the current that does actual work (and that, over time, is metered as consumed energy, i.e. watt-hours). It is important to remember this discrepancy between watts delivered and volt-amperes being consumed when selecting circuit components. This discrepancy between watts and volt-amperes is reflected in the power factor of the motor (See Glossary).

The amperage drawn from the power line depends on the horsepower delivered by the motor—whether it is overloaded or underloaded. Your cost for the power is based on watts consumed, but you must provide wire size in proportion to the amperes. When the motor is first turned on it momentarily consumes several times its rated current. After it comes up to speed but is permitted to idle, delivering no load, it consumes about half its rated current. Rated current is consumed when the motor is delivering its rated horsepower, and more current is consumed if it is overloaded.

TYPES OF MOTORS

Choose the type of motor that is suitable for the job and the available power source. Described here are common ac motors. Features to consider are size, starting capacity, initial cost, operating cost, and ease of maintenance.

The direction of rotation can be changed on ac motors. On a repulsion-induction motor, shift the position of the brushes. On other types of ac motors, reverse two of the wires coming from the starting winding inside the motor. If the motor must be reversed often, a special reversing switch can be installed for the purpose.

Single-phase motors   The motors most commonly found in homes and on farms are called single-phase motors because they operate on the usual 120/240-volt, single-phase current. They are not usually available in sizes larger than 7 1/2 hp although a few larger ones are made. The three types described in this discussion operate only on single-phase ac.

Larger single-phase motors are often dual-voltage, meaning they are designed to be operated at either of two voltages, for example 120 or 240 volts. The motor has four leads. Connected one way the motor operates at 120 volts; connected the other way it operates at 240 volts. See Fig. 16-1.

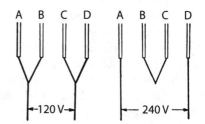

If there is a choice, always operate your motor at the higher voltage. At 240 volts it will consume only half as many amperes as at 120 volts. With any given

Fig. 16-1 By reconnecting the leads, the motor can be used on either 120 or 240 volts.

wire size the voltage drop will be only one-fourth as great (measured as a percentage) on the higher voltage than it would be on the lower voltage.

Split-phase motors   This is a simple type of motor, which makes it relatively trouble free; there are no brushes, no commutator. It is available only in sizes of 1/3 hp and smaller. It draws a very heavy amperage while starting. Once up to full speed, the split-phase motor develops just as much power as any other type of motor, but it is not able to start heavy loads. Therefore do not use it to drive any machine that is hard to start, such as a deep-well pump, or an air compressor that has to start against compression. Use it on any machine that is easy to start, or on one where the load is thrown on after the machine is up to full speed. It is entirely suitable for washing machines, grinders, saws, lathes, and general utility use.

Capacitor motors   This is similar to the split-phase type, with the addition of a capacitor or a condenser that enables it to start much harder loads. There are several grades of capacitor-type motors available, ranging from the home-workshop type which starts loads from 1 1/2 to 2 times as heavy as the split-phase, to the heavy-duty type which will start almost any type of load. Capacitor motors usually are more efficient than split-phase, using fewer watts per horsepower. The amperage consumed while starting is usually less than half that of the split-phase type. Capacitor motors are commonly used only in sizes up to 10 hp.

Repulsion-induction motors   This type of motor, properly called the "repulsion start, induction run" motor, is commonly called a repulsion-induction or R-I motor. Available in sizes up to 10 hp, it has a very high starting capacity and should be used for heavier jobs—it will "break loose" almost any kind of hard-starting machine. The starting current is the lowest of all the single-phase types of motors. These motors have a commutator connected to the windings on the rotor which allows for precise positioning of the rotor poles (by moving the brush positions) in relation to the field poles to produce maximum starting torque. Commutators and brushes require skilled maintenance. With the increased availability of three-phase power, use of R-I motors is decreasing.

Three-phase motors   These motors are the simplest and most trouble-free type made and, as the name implies, operate only on three-phase ac. Three-phase motors in sizes 1/2 hp and larger cost less than any other type, so use them if you have three-phase current available. Large motors (over 10 hp) are only available in this form. Reversing any two phase connections reverses the motor.

LARGE MOTORS ON FARMS

On farms, motors of 10 hp, 25 hp and even larger are required. But many farms have only a single-phase, three-wire, 120/240-volt service. These services usually only require one high voltage line to the farm and one transformer. Single-phase motors 5 hp and larger require high starting currents that may disrupt the voltage supplied to other customers when they start.

Before buying even a 5-hp single-phase motor, check with your power supplier to see whether their service standards allow you to operate such a motor. If you are fortunate enough to have three-phase service, your problems are solved. Use three-phase motors, which are simpler in construction so they cost considerably less than single-phase motors and rarely have a service problem. Note: If you have three-phase service, you may have three-phase power at 240 volts available in addition to the usual 120/240-volt, single-phase for lighting, appliances, and other small loads.

Three-phase distribution practices are increasingly moving in the direction of 208Y/120-volt systems, in which the voltages between the neutral and all three ungrounded conductors are equally 120 volts. The older 240-volt systems have 120 volts to two of the ungrounded conductors from the grounded conductor, and 208 volts to the third. This makes it more difficult for the power suppliers to balance the 120-volt loads on their systems, and it creates a potential hazard if the property owner inadvertently connects a 120-volt load incorrectly at 208 volts.

PHASE CONVERTERS

What can you do if you need larger motors but have only the usual single-phase, 120/240-volt service? A phase converter permits three-phase motors to be operated on single-phase lines. The phase converter changes the single-phase power into a modified three-phase power that will operate ordinary three-phase motors and at the same time greatly reduce the starting current required. In other words, when operating a three-phase motor with the help of a phase converter, the same single-phase line and transformer that would barely start a 5-hp single-phase motor will start a 7½-hp or possibly even 10-hp three-phase motor, and a line and transformer that would handle a 10-hp single-phase motor (if such a motor could be found) would probably handle a 15-hp or 20-hp three-phase motor. The *NEC* requirements for phase converters are in Article 455. Phase converters are expensive and may require power supplier approval, but their cost is partially offset by the lower cost of three-phase motors and the increased labor efficiency gained when larger machinery can be used.

Two types of converters   The *static type* of phase converter has no moving parts except relays. It must be matched in size and type with the one particular motor to be used with it; generally, there must be one converter for each motor. The *rotating type* of converter looks like a motor, but can't be used as a motor. Two 240-volt, single-phase wires run into the converter; three 3-phase wires run out of it. Usually several motors can be used at the same time. The total horsepower of all the motors in operation at the same time can be at least double the horsepower rating of the converter. Thus, if you buy a converter rated at 15 hp, you can use any number of three-phase motors totaling not over 30 to 40 hp, but the largest may not be more than 15 hp—the rating of the converter. The converter must be started first, then the motors, starting with the largest and then the smaller ones.

Required horsepower ratings   Some words of caution are in order. A three-phase motor of any given horsepower rating will not start as heavy a load when operated from a phase converter as it will when operated from a true three-phase line. For that reason, it is often necessary to use a motor one size larger than is necessary for the running load. This does not significantly increase the power required to run the motor once it is started. The converter must have a horsepower rating at least as large as that of the largest motor.

The voltage delivered by the converter varies with the load on it. If no motor is connected to the converter, the three-phase voltage supplied by it is much higher than the input voltage of 240 volts. Do not run the converter for significant periods without operating motors at the same time or it will be damaged by its own high voltage. Do not operate only a small motor from a converter rated at a much higher horsepower because the high voltage will damage the motor or reduce its life. It is good practice to make sure the total horsepower of all the motors operating at one time is at least half the horsepower rating of the converter.

Variable-speed drives can be used for phase transformations   Variable-speed drives, customarily considered for speed regulation, work by taking the incoming ac supply and rectifying it to dc. Then the drive rebuilds an ac waveform of the required frequency to produce the desired speed. Since it is working from a dc bus and because all the controls are electronic, most drives can build a three-phase supply as easily as they can build a single-phase supply (although their horsepower rating will be derated). This approach is gradually supplanting the role of phase converters. For example, it is common to find submersible well pump controls working on this principle, even in residential applications. The controller is connected to a 240V single-phase supply, and it generates a three-phase supply for the motor at the bottom of the well shaft. Three-phase motors are much simpler to operate and don't have some of the maintenance problems that single-phase motors do.

MOTOR CIRCUIT REQUIREMENTS

Separate circuits are recommended for individual motors that are not part of an appliance if they are of more than ⅛ hp, whether directly connected or belt driven. Use Table 5–2, adapted from *NEC* Table 430.148, to determine the motor's amperage. Every motor must have a disconnecting means, a controller to start and stop it, short-circuit and ground-fault protection, and motor overload protection in case of overload or failure to start. Several of these are often combined.

Disconnecting means   Every motor must have at least one disconnecting means capable of completely isolating the motor from all ungrounded conductors so it can be maintained safely. The *NEC* rules on motor disconnecting means fall into two categories. First, the disconnecting means must be suitable to handle this duty, and second, it must be located where those who may need to operate it will be able to do so promptly. That concept is more fully explained under the heading "'In sight from' requirements" at the end of this discussion of motor circuit requirements. Acceptable motor disconnecting means generally used on projects within the scope of this book include the following devices:

Motor-circuit switch—A switch, rated in horsepower, capable of interrupting the maximum operating current of a motor of the same horsepower rating as the switch at the rated voltage. The device must be listed to qualify.

Circuit breaker—A circuit breaker, in the panel or separately mounted, qualifies.

Molded case switch—A nonautomatic circuit breaker. Molded case switches contain the switching mechanism and manually operable handle of circuit breakers, but no thermal or magnetic sensing mechanism that would cause an automatic trip.

Manual motor controller additionally marked SUITABLE AS MOTOR DISCONNECT— Figure 16–2 shows some manual motor controllers. Even though they are controlled manually and have OFF and ON positions, they do not qualify as disconnecting means without meeting additional qualifications because they differ in robustness of construction from motor circuit switches. The *NEC* allows them in two circumstances. The first, covering small motors, allows them to be used as disconnects for motors of 2 hp or less, just as snap switches (described later in this list). The second,

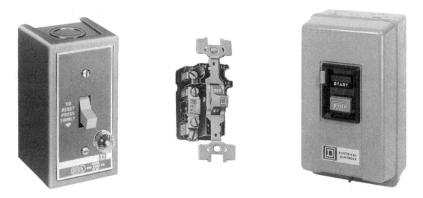

Fig. 16-2 Controls of these types are "manual motor controllers." They are used to start and stop a motor; they also contain motor overload devices. If marked SUITABLE AS MOTOR DISCONNECT they can additionally function as the required disconnecting means. *(Square D Company)*

covering larger motors, allows them to be used as disconnects if they are on the load side of the final branch-circuit short-circuit and ground-fault protective device. In either case, their horsepower rating must not be less than the motor.

General-use switch—A switch intended for use in general distribution and branch circuits. It is rated in amperes, and is capable of interrupting its rated current at its rated voltage. See Fig. 16-3 for an example. Its ampere rating must be not less than twice the full-load current rating of the motor. It generally cannot be used for a motor larger than 2 hp, unless it additionally qualifies as a motor-circuit switch, as described earlier. Use a switch with one fuse for a 120-volt motor, and two fuses for a 240-volt motor.

General-use snap switch—A form of general-use switch constructed so that it can be installed in flush device boxes or on outlet box covers, or otherwise used in conjunction with wiring systems recognized by the *NEC*. (In other words, these are the ordinary switches used in controlling lights in ordinary house wiring.) They are for ac motors only. To qualify, the switch must be rated ac-only (general-use ac-dc snap switches are not acceptable) and the motor full-load current must not exceed 80 percent of the ampere rating of the switch.

Plug and receptacle—If the motor is portable, the plug on the cord is sufficient if the motor rating does not exceed the established horsepower

Fig. 16-3 A switch of this type, or a larger one with cartridge fuses, may be used with small motors. If the switch has two fuses, it is for a motor operated at 240 volts; if it has only one fuse, it is for a motor operated at 120 volts. *(Square D Company)*

ratings for the configuration. For example, a 125-volt, 15-amp receptacle can be used with up to a ½-hp motor; a 125-volt, 20-amp receptacle is good for 1 hp; a 250-volt, 15-amp receptacle is rated to 1½ hp; and the 20-amp variety is rated for 2 hp.

Controllers   A controller is any device used to start and stop a motor. It is part of the machine on refrigerators, pumps, and other equipment with automatically started motors. On manually started motors it can be a circuit breaker or switch, but is usually what is called a motor starter, as shown in Fig. 16–2. The enclosure for the starter also contains motor overload devices, which are discussed under the next heading. Controllers for motors over 2 hp must have a horsepower rating equal to or more than the horsepower rating of the motor. Use the smaller starter in the illustration for fractional-horsepower motors; it has a manual switch to start and stop the motor. For bigger motors, use the larger starter shown at the right. It has pushbuttons to start and stop the motor. Larger starters have pushbuttons in a separate case arranged to actuate a magnetic switch (usually called a "contactor"), allowing the motor to be electrically controlled from a distance and from multiple locations or by automatic means, such as through a pressure switch.

Motor overload devices   Motors must be protected by overload devices to prevent burnout and risk of fire resulting from extended overload.

It takes many more amperes to start a motor than to keep it running at full speed at its rated horsepower. And when a motor is overloaded, it consumes more amperes than while delivering its rated horsepower. A motor will not be damaged by current considerably larger than normal flowing through it for a short time, as at startup or during a momentary overload. But it will burn out if more than normal current flows through it for a considerable time.

A motor overload device permits the high starting current to flow for a short time, but disconnects the motor if current due to overload (or failure to start) flows through it for a considerable time. Overload devices are permitted to be separate, but in practice they are usually included in the same enclosure with the starter.

Overload devices include "heaters" and are rated in amperes. When the starter is installed, select an overload device heater on the basis of the full-load ampere rating on the nameplate of the motor. Overload devices are often integral with (built into) the motor in the case of small motors, also with many larger motors if they are part of automatically started equipment such as air-conditioning units. Whether integral or separate, if the overload device stops the motor, correct the condition that led to the overload. Let the motor and the overload device cool off, then reset manually.

Some overload devices are automatically resetting, but they must not be used where the unexpected restarting of the motor (for example when powering a table saw) could result in injury.

Motor branch-circuit short-circuit and ground-fault protection   Motor overload devices, whether built into the motor or installed on the starter, are not capable of interrupting the high amperage that can arise instantaneously in case of a short circuit or ground fault that might occur in the motor circuit or in the motor. The branch circuit as well as the controller and the motor must be protected by fuses or circuit breakers against such shorts or ground faults.

The wires in a motor branch circuit must have an ampacity of at least 125 percent of the full-load motor current so they won't be damaged if the motor is overloaded. (Overload devices permit up to about 25 percent overload current for a considerable time before stopping the motor.) For motors of the kind typically installed in homes and on farms, the *NEC* permits a breaker to have an ampere rating up to 250 percent of the full-load current of the motor; if time-delay fuses are used, their ampere rating must not exceed 175 percent of the full-load current. If these values seem to contradict the basic requirement that overcurrent devices may not have an ampere rating larger than the ampacity of the wire being protected, bear in mind that a motor circuit is a special case in which the overcurrent device protects only against short circuits or grounds. Protection against lower values of overcurrent (overload, or failure to start) is provided by the motor overload device discussed under the previous heading. Use the smallest rating that will permit the motor to start and operate properly. Fuses that are not of the time-delay type should not be used.

"In sight from" requirements   A motor disconnecting means must be in sight from the motor controller; there are no exceptions that normally apply to this rule. In addition there must be a disconnect in sight from the motor and its driven machinery. The *NEC* defines "in sight from" as being visible and not over 50 feet from the specified location. If you can see one component from the other but they are more than 50 feet apart, they are not "in sight from" each other. This concept, which originated in the *NEC* motor article, is formally defined in *NEC* Article 100 for use throughout the *NEC*.

If the controller disconnect is not in sight from the motor and its driven machinery, you must install an additional disconnecting means that is in sight from the motor and its driven machinery. This additional disconnecting means must meet the requirements for the disconnecting means already discussed, but if it is a switch it need not have fuses. There is an exception that allows the in-sight disconnect to be omitted if the disconnecting means for the controller can be individually locked in the open position; however, effective with the 2002 *NEC* that exception is limited to installations where the additional disconnect would introduce additional hazards or would be impracticable. For example, it would be plainly impracticable to place a disconnect 50 feet down a well shaft to be "in sight" (not over 50 feet distant) from a submersible pump motor 100 feet down the same shaft.

INSTALLING THE MOTOR

Observe the requirement outlined above for disconnecting means, controller, and overload units. Use wire with an ampacity of at least 125 percent of the full-load current of the motor. For long circuits use wire larger than required by *NEC* in order to avoid excessive voltage drop. If the circuit to the motor is long, and you use wire that is too small, you may have only 100 volts at the motor during the starting period instead of nearly 120 volts. The motor may not start, or if it does it will deliver

Table 16-1   ALLOWABLE DISTANCE (IN FEET) FROM SERVICE EQUIPMENT TO SINGLE-PHASE MOTOR FOR DIFFERENT WIRE SIZES

Hp		14 AWG	12 AWG	10 AWG	8 AWG	6 AWG	4 AWG	2 AWG	1/0 AWG	3/0 AWG
¼	120 volts	55	90	140	225	360	575	900	1,500	2,300
⅓		45	75	115	180	300	450	725	1,200	1,800
½		35	55	85	140	220	350	550	850	1,400
¾			40	60	100	150	250	400	600	1,000
1			35	50	85	130	200	325	525	850
¼	240 volts	220	360	560	900	1,450	2,300	3,600		
⅓		180	300	460	720	1,200	1,600	2,900		
½		140	220	340	560	875	1,400	2,200		
¾		100	160	240	400	600	1,000	1,600	2,400	
1		85	140	200	340	525	800	1,300	2,100	
1½	240 volts	70	110	160	280	400	675	1,100	1,700	
2		60	90	140	230	350	550	900	1,400	2,200
3				100	160	250	400	650	1,000	1,600
5					100	160	250	400	650	1,000
7½						110	175	275	450	700

The wires shown in the table are compromises. The voltage drop while starting (when the amperage is several times greater than when the motor is running) will be from 3 to 7½ percent depending on the type of motor, how hard the load is to start, and other factors. While running, the drop will be about 1½ percent. This is less than normally recommended for non-motor loads and is the result of wire sizes large enough to keep the drop within reasonable limits while starting.

significantly less than its rated horsepower. Table 16-1 shows the wire size to use at various distances from service equipment to motor, assuming the wire has an ampacity of at least 125 percent of the full-load current shown on the nameplate of the motor. All distances are one-way. The table is for single-phase motors only.

Chapter 17
FARM WIRING

FARMHOUSES HAVE ABOUT THE SAME power requirements as city houses and may be wired as any other house. It is the rest of the farm that makes up the main part of the load. This includes all sorts of motor-driven machinery: water pumps and water heaters; irrigation equipment; milking machines, milk coolers, sterilizers, and similar dairy equipment; extra lighting for forced egg production, incubators and brooders; crop-drying equipment; heaters to keep water for livestock or poultry from freezing, and so on.

To provide enough power to operate such equipment, you must use large wires to carry the heavy amperages without undue voltage drop. The large loads require heavy service equipment and many circuits. This chapter looks at the wiring considerations encountered in providing the power needed to carry on the business of farming. Problems with large motors on farms are discussed on pages 179–180.

Wiring issues and procedures specific to farming are collected in this chapter for convenient reference, but it is important to study the entire book in order to gain understanding of the principles involved.

On practically all farms, the power supplier's wires end on a pole in the farmyard, which is where this chapter begins. On the pole are the meter and often a switch or circuit breaker to disconnect the entire installation. The wires are grounded at the pole. From the top of the pole, sets of wires run to the house, to the barn, and to the other buildings to be served. At each building there is a service entrance without a meter. Underground cables are being used more and more instead of overhead wiring. The type of cable you use depends on the purpose of the building. The remainder of the chapter discusses electrical requirements for farms predominately in terms of the requirements of *NEC* Article 547, which applies to at least portions of most but not all agricultural buildings.

LOCATING THE METER POLE ("DISTRIBUTION POINT" IN *NEC* TERMINOLOGY)

Why is there a pole? Why not run the wires to the house and from there to the other buildings as was done when farms were first being wired? Because that would lead to very large wires to carry the total load involved; very large service equipment in the house; expensive wiring to avoid voltage drop—which is wasted power; a cluttered farmyard, and many other complications that can be avoided by installing a meter pole. The meter pole should be located, if possible, near the buildings with the heaviest loads in order to reduce voltage drop and the length of large-capacity feeders.

Overhead or underground? A farm wired using overhead wires will have a large number of wires running all over the place. This gives a very untidy appearance and invites problems of various kinds. Wires to low buildings are subject to damage by vehicles and machinery passing under them. Having many overhead wires on an isolated farm increases vulnerability to lightning. In northern climates where sleet storms are common, wires can break from the weight of accumulated ice; a broken wire on the ground is dangerous. Underground wires cost a little more but prevent these problems. However, underground cable faults are comparatively difficult to find and repair.

If you are going to use overhead wiring, consider wire size carefully. Select a wire size big enough to carry its load in amperes without excessive voltage drop. Make sure it is strong enough for the length of the span, which means that sometimes you must use wires larger than otherwise necessary for the load involved. Review the voltage drop topic in Chapter 4.

Basic construction at pole Usually, three wires come from the power supplier's transformer and end at the top of the pole. The usual practice is to use triplex cable (See Fig. 8–19 for an example), but separate wires are shown in the drawing (Fig. 17–1) to make it easier to follow. Check with your power supplier regarding how they intend to supply the pole. The neutral is always grounded at the pole.

Figure 17–1 shows a modern farm distribution point. The disconnecting means may be provided by the utility or by the owner, depending on local practice. The *NEC* classifies this device as a "site isolating device" to distinguish it from a service disconnect. Since it is not an actual service disconnect, it follows that the wiring that leaves this device still has the status of service conductors and must meet the wiring procedures and clearance requirements described under this topic in Chapter 8. Note that although the switch is at the top of the pole, it can be operated from a readily accessible point through the permanently installed linkage shown in Fig. 17–1. In addition, you must install a grounding electrode conductor at this point and run it from the neutral block of the switch to a suitable electrode at the pole base.

From the pole top, go to the buildings that need to be supplied from this point. As a general rule, the farmhouse can be supplied by a three-wire service, with its neutral regrounded at the home just as if the utility had made a direct termination. The farmhouse must not, however, share a common grounding return path with the barn. If it does, as in the case of a common metallic water piping system, the house (1) has to be supplied with a four-wire service, and (2) have all instances of electrical contact between the neutral and the local equipment grounding system removed.

Although the barn, arguably, could also be wired like the house (three-wire), a three-wire hookup would mean that the neutral and the equipment grounding system in the barn would be bonded together at the barn disconnect. That in turn would mean the neutral, in the process of carrying current across its own resistance, would constantly elevate the voltage to ground of all barn equipment by some finite amount relative to local ground, especially as it appears to farm animals where they

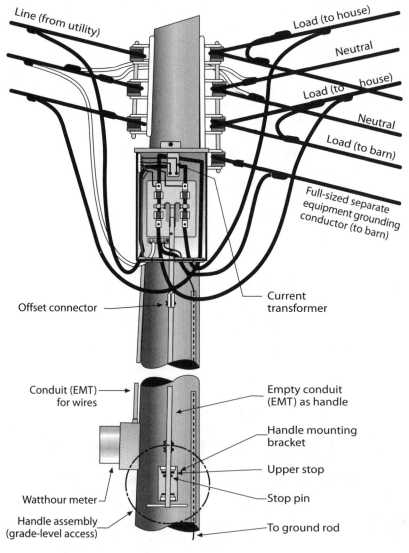

Fig. 17–1 Pole top switch and remote metering as covered in *NEC* 547.9(A). Leave at least half the circumference of the pole clear to allow line workers and repair workers to climb the pole without trouble.

stand. The feet of livestock, being in close contact with moisture, urine, and other farm chemicals, are conductively rather well coupled to local earth. Most livestock are much more sensitive to voltage gradients than people are. A potential difference in the range of a fraction of a volt can take a cow out of milk production, which no farmer can afford.

Four-wire distribution may be required   The *NEC* addresses this in two ways. First, it establishes the unique rules on farm distributions being covered here. Second, it establishes an equipotential plane for these environments, discussed later in this chapter. The service to the barn is normally wired four-wire, but there is an additional condition attached to the four-wire scheme that is unique to agricultural buildings. The separate equipment grounding conductor must be fully sized. That is, if the run to the barn is 3/0 AWG copper for a 200-amp disconnect, and the neutral is 1/0 AWG copper (both sized on the basis of load), the equipment grounding conductor is not 6 AWG as normally required by *NEC* Table 250.122; nor is it 4 AWG, the size for a grounding conductor on the supply side of a service using 3/0 AWG wires; nor is it 1/0 AWG, the size of the neutral. It must not be smaller than the largest ungrounded line conductor, or 3/0 AWG. When this wire arrives at the barn, it must arrive at a local distribution with the neutral completely divorced from any local electrodes or equipment surfaces requiring grounding.

Installing the meter socket   The meter socket is sometimes furnished by the power supplier but installed by the contractor. In other localities it is supplied by the contractor. As shown in Fig. 17–1, the meter socket is not connected across the line, but instead is connected through a current transformer (usually abbreviated CT), discussed under its own heading later in this chapter. Mount it securely at a height required by the local authorities; this is usually 4 to 6 ft above ground level.

In the case of a small service, the actual line and load conductors could come down the pole to a meter socket connected across the line. In this case the usual procedure is to wire the service disconnect immediately below the meter socket. Run service wires from the bottom jaws of the meter socket to the line terminals of a switch or circuit breaker. Then run the load wires from the load terminals of the service disconnect back up to the point of beginning at the pole top. Some power suppliers allow all the wires in one conduit ("single stack") and some prefer two conduits ("double stack"). Pass the load wires, unbroken, straight through the meter socket on their way back up. Combination meter sockets and main disconnects, usually using a circuit breaker, are commonly available and simplify this process. In sizing the service and the load wires, be aware that if you install them in a common raceway (the usual case) you must count the four hot wires as current carrying. That number of current-carrying wires run together requires their ampacities (from Table 4–1) to be derated by multiplying by a factor of 0.8. Large farm loads, however, almost always use the pole top switch and CT shown in Fig. 17–1.

Insulators on pole   Near the top of the pole, install insulator racks of the general type shown in Fig. 17–6. Provide one rack for the incoming power wires and one rack for each set of wires running from the pole to various buildings. Remember that there is great strain on the wires under heavy winds or in case of heavy icing in northern areas. Anchor the racks with heavy lag screws. Better yet, use at least one through-bolt, all the way through the pole, for each rack. Note the fourth insulator on the pole for the equipment grounding conductor run to the barn.

Ground at meter pole   For a small service using a meter connected directly across the line, the neutral wire always runs from the top of the pole through the conduit to the center terminal of the meter socket. The neutral is not necessary for proper operation of the meter, but grounds the socket. At one time it was standard practice to run the ground wire from the meter socket (out of the bottom hub) to ground, but experience has shown that better protection against lightning is obtained if the ground wire is run outside the conduit. Run it from the neutral at the top of the pole directly to the ground rod at the bottom of the pole, terminated as shown in Fig. 17-2. It is usually run tucked in alongside the conduit as far as it goes, then to ground. In some localities it is stapled to the pole on the side opposite the conduit.

For larger services using pole-top disconnects and CTs, just run the ground wire straight down the pole. Protect it from damage, particularly along the lower end of the pole, with U-shaped heavy plastic coverings, or run it in rigid nonmetallic conduit. If you use steel conduit, make sure both the top and bottom ends are bonded to the enclosed ground wire as described in Chapter 8 (under "Bonding at the service entrance") or the ground wire will lose a good portion of its effectiveness. Complete the installation using the procedure covered later in this chapter under "Burying the ground rod." How to determine the size of the ground wire is discussed under its own heading in Chapter 8; remember that if the ground wire runs to a rod, pipe, or plate electrode, such as a ground rod, it never needs to be larger than 6 AWG.

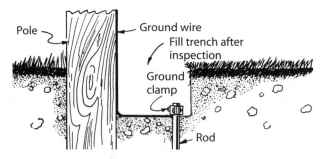

Fig. 17-2 The top of the usual 8-ft ground rod should end up below the surface of the ground. Fill the trench after inspection. If the ground rod is copper, use only a copper clamp. Per listing restrictions, insist on a DIRECT BURIAL or DB marking on the clamp.

DETERMINING SIZE OF WIRES AND SERVICE

Wires of the proper size must run from the yard pole to individual buildings. At each building there must be a service entrance as discussed in Chapter 8, except the entrance will not have a meter. The service entrances must have the correct ratings, and correct ampacities must be determined for the wires running from the pole to each building and for the wires on the meter pole.

The *NEC* provides a specific method for determining the ampacity of the wires running from the pole to any farm building with two or more circuits, and for determining the rating of the service equipment at that building. The minimum size of wires on the pole is based on the total load on all wires between pole and individual buildings.

The wire sizes determined by using the formula that follows are the minimum permissible by *NEC*. You would be wise to install larger sizes to allow for future expansion. Number 10 AWG wire is acceptable for spans up to 50 feet, and 8 AWG for longer spans. For very long spans, use an extra pole. If the wires are installed in northern areas on a hot summer day, remember that a copper wire 100 feet long will be a couple of inches shorter next winter when the temperature is below zero. Leave a little slack so insulators will not be pulled off buildings during winter.

It is assumed that the building will have a three-wire, 120/240-volt service, so the total amperage *at 240 volts* must be determined. For motors, use the amperage shown in Table 5–2. For all other loads, start with the wattage. For lights, you can determine the total watts from the lamp size you intend to use. For fluorescent lights, add about 15 percent to the wattage because the watts rating of fluorescents defines only the power consumed by the lamp itself; the ballast adds from 10 to 20 percent. For receptacles, allowing 200 watts for each will probably be on the safe side since they will not all be used at the same time. Then divide the total watts by 240 and you will have the amperage at 240 volts.

Note: Suppose in a building you have six 120-volt, 15-amp circuits for lights and receptacles—theoretically making a total of 90 amps (6×15) at 120 volts, or 45 amps at 240 volts. But the circuits won't all be loaded to capacity at the same time, so instead of the 45 amps, use the total as determined by the preceding paragraph.

Minimum service for each farm building   *NEC* 220.103 and 220.102 show how to compute the load for service conductors and service equipment for each farm building (except the house, which is computed as any other house) and the total farm load. For each building to which wires run from the pole (except the house—see "What Size Service" in Chapter 8), first determine the amperage at 240 volts of all loads having any likelihood of operating *at the same time*. Enter the amperage at *a*, then proceed with steps *b, c, d, e* and *f*.

a.	Amperage at 240 volts of all connected loads likely to be operating at the same time, including motors, if any	_____ amps
b.	If *a* includes the largest motor in the building, add here 25% of the amperage of that motor (if two motors are the same size, consider one of them the largest)	_____ amps
c.	If *a* does not include the largest motor, show here 125% of the amperage of that motor	_____ amps
d.	Add *a* to *b* or *c*, whichever is applicable	_____ amps

e. Amperage at 240 volts of all other connected loads in the   _____ amps
  building

f. Total of *d* + *e*   _____ amps

Using your results from above, select the appropriate case from the following choices to find the minimum size of service for each individual building.

▦ If *f* is 30 or less, and if there are *not more than two* circuits, use a 30-amp switch and 8 AWG wire (8 AWG is *NEC* minimum). If there are *three or more* circuits, use a 60-amp switch and 6 AWG wire.

▦ If *f* is over 30 but under 60, use a 60-amp switch and 6 AWG wire.

▦ If *d* is less than 60 and *f* is over 60, start with *f*. Add together 100% of the first 60 amps plus 50% of the next 60 amps plus 25% of the remainder. For example, if *f* is 140, add together 60 plus 30 (50% of the next 60 amps) plus 5 (25% of the remaining 20 amps) for a total of 60 + 30 + 5 or 95 amps. Use 100-amp switch and wire with ampacity of 95 amps or more.

▦ If *d* is over 60 amps, start with 100% of *d*. Then add 50% of the first 60 amps of *e*, plus 25% of the remainder of *e*. For example, if *d* is 75 amps and *e* is 100 amps, start with the 75 of *d*, add 30 (50% of the first 60 of *e*), and add 10 (25% of the remaining 40 of *e*) for a total of 75 + 30 + 10 or 115 amps. Use a switch or breaker of not less than 115-amp rating and wire with corresponding ampacity.

Calculating minimum size of wires on pole The calculations above determined the size of the service to each building. Now let us determine the size of wires on the pole from the top of the pole where they are connected to the power supplier's wires, down to the meter, and back up to the top of the pole where they are connected to the wires leading to individual buildings. The wires on the pole and the main disconnect must be large enough to handle all the loads calculated above. The total number you enter in step 7 of the following calculation is the minimum rating of the switch or breaker (if used) at the pole, and the minimum ampacity of the wires on the pole.

Proceed as follows using the amperages determined in the previous calculation:

1. Highest of all the amperages for _____ amps at 100% = _____ amps
  an individual building (excluding
  the house) as determined in
  previous calculation

2. Second highest amperage   _____ amps at 75% = _____ amps

3. Third highest amperage   _____ amps at 65% = _____ amps

4. Total of all other buildings   _____ amps at 50% = _____ amps

5. Total all above   _____ amps

6. Add the house, as figured   _____ amps
  in Chapter 8

7. Total of 5 + 6   _____ amps

Note 1   If two or more buildings have the same function, consider them as one building for the purpose above. For example, if there are two brooder houses requiring 45 and 60 amps respectively, consider them as a single building requiring 105 amps. If no other building requires more than 105 amps, enter 105 amps in step 1 above.

Note 2   In listing the amperage for any one building, the amperage to use is the calculated amperage, not the rating of the switch used. For example, if for any building you determined a minimum of 35 amps but you use a 60-amp switch (because there is no size between 30 amps and 60 amps), use 35 amps not 60 amps.

Use current transformers for modern high-capacity installations   If the service at the pole is rated at 200 amps or more (line 7 of the preceding calculation), it means very large wires running from the top of the pole down to the meter and back again to the top. That is both expensive and clumsy. It is not necessary to run such large wires down to the meter—instead use a current transformer (CT).

An ordinary transformer changes the voltage in its primary to a different voltage in its secondary. In a current transformer, the current flowing in its primary is reduced to a much lower current in its secondary. Most current transformers are designed so that, when properly installed, the current in the secondary will never be more than 5 amps. A typical current transformer, shown in Fig. 17–3, has the shape of a doughnut 4 to 6 inches in diameter. It has only a single winding: the secondary. The load wire (in which the current is to be measured) is run through the hole of the doughnut and becomes the primary. Assuming that the transformer has a 200:5 ratio, for use with a 200-amp service, the current in the secondary will be 5/200 of the current in the primary. If the current in the primary is 200 amps, 5 amps will flow in the secondary; if it is 100 amps in the primary, 2½ amps will flow in the secondary. For a 400-amp service, the transformer would have a 400:5 ratio.

One current transformer is installed at the top of the pole, with the two hot wires running through the hole. Four small 14 AWG wires run from the top of the pole to the meter (which must be of the type suitable for use with a current transformer): two from the

Fig. 17–3 A current transformer.

secondary of the current transformer and two from the hot wires, for the voltage. The meter operates on a total of not more than 5 amps but the dials of the meter will show the actual kilowatt-hours consumed. A wiring diagram is shown in Fig. 17–4.

Caution: The secondary terminals of the transformer must always be short-circuited while any current is flowing in its primary. If connected to a kilowatthour meter, the meter constitutes a short circuit. As purchased, the transformer will probably have a short-circuiting bar across its secondary terminals; this must not be removed until the transformer is connected to the meter. If you were to touch the terminals of such a meter that is not short-circuited, you would find a dangerous voltage of many thousands of volts.

In an installation using a current transformer at the top of the pole with a site-isolating switch, the CT is often installed in the switch enclosure as shown in Fig. 17–1. Such switches are also available in the double-throw type, as required if a standby generating plant is installed for use during periods of power failure.

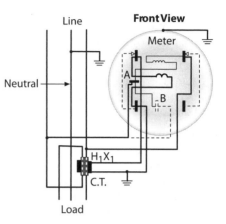

Fig. 17–4 Wiring diagram showing one transformer used on a 120/240-volt service.

INSTALLING OVERHEAD WIRING

Triplex cable is shown in Fig. 8–19; it consists of two insulated wires wrapped spirally around a strong, bare neutral wire. It requires only one insulator for support. One triplex cable is usually considered preferable to three separate wires. It is frequently used for overhead spans to farm buildings generally, and it is available as "quadruplex" cable with four wires, allowing it to be used on four-wire distributions as covered previously. The *NEC* covers it as Article 396, "Messenger Supported Wiring," and power suppliers have been using it for service drops for years.

Required clearance of overhead service wires from grade is given in *NEC* 230.24(B). For feeder wires from building to building on the farm, the clearances are given in *NEC* 225.18 as follows:

■ 10 feet above finished grade, sidewalks, or any platform or projection from which wires might be reached

■ 12 feet above residential property and driveways and commercial areas not subject to truck traffic

■ 18 feet over commercial or farm properties subject to truck traffic

For conductors over 600 volts, the National Electrical Safety Code, ANSI C2-1997 is referenced. Overhead transmission and distribution lines of the power supplier often abut or cross farm property. Clearances may have to be increased on the farm due to use of mobile equipment that is unusually high, such as grain augurs and conveyors. Dangerous situations can arise when working under an energized electric line with well-digging equipment, grain probes, irrigation pipe, and well pipe. Proper layout of the overhead electrical distribution system with relation to the structures and proximity to the power lines can save lives by minimizing exposure to potential shock. Consult with a representative of your local power supplier for advice on minimizing hazards.

INSTALLING UNDERGROUND WIRING

It is becoming more common for service entrances, and also for wires between buildings, to be put underground, producing neater installations. Underground wiring greatly reduces danger from lightning and eliminates the problem of long spans coming down under ice loads in northern areas.

Underground wiring uses either conduit or special cables designed for direct burial. The cables have special moisture-resistant insulation on the conductors, with a very tough outer layer. The *NEC* recognizes two types, which are described below. They are available in both single-conductor and multi-conductor configurations.

USE and UF cable   The most common cable for underground wiring is Type USE (underground service entrance), which has been used for many years. It is shown in Fig. 17–5. Another type, Type UF (underground feeder), may be used as an underground feeder, but must be *protected by fuses or breaker at the source*; therefore it must not be used as a service entrance conductor. However, on farms, Type UF could be used on underground runs from pole to building if protected by fuses or breakers.

In the multi-conductor kind, Type UF is substantially the same *in appearance* as Type NMC shown in Fig. 4–5B. Nevertheless, they are two different kinds of cable.

Fig. 17–5 Type USE cable is designed to be buried directly in the ground without further protection. It is available also as 2-wire or 3-wire cable.

Type NMC may be used in dry, moist, or damp locations, but not for direct burial underground. Type UF may be buried directly in the ground and may also be used wherever Types NM and NMC may be used.

Minimum depth   Use common sense in installing any kind of underground cable. *NEC* 300.5 covers the minimum depth to which it must be buried. For *residential* installations the minimum depth is 12 inches, provided the cable is protected by a breaker or fuses rated at 20 amps or less, and is protected by a GFCI. If not so protected, the cable must be buried a minimum of 24 inches, which is also the minimum depth required for all nonresidential installations. If you use individual wires, keep them close together. In locations where the cable might be disturbed (as where it crosses roadways, or where it crosses cultivated areas where there may be future digging), bury it deeper and lay a board or similar protective material above the cable before filling the trench. In general, a 2-inch thick run of concrete placed in the trench above the underground wiring allows a reduction in cover by 6 inches, although this is not the case in areas subject to vehicular access. Do not stretch the cable tightly; lay it so it curves snakelike from side to side. To avoid damage to the conductors, screen any gravel or rocks out of the material used for backfilling the trench, or better yet use clean river sand.

If you are using a raceway, the burial depths are not as great, and depend on the raceway chosen. Heavy-wall steel conduits require only 6 inches of cover and

PVC 18 inches. These distances become consistent for all wiring methods, even for direct-burial cables, at 18 inches under residential driveways and parking areas and 24 inches under public roadways and parking lots. Note that the cover dimension is measured between the top of the raceway and grade level, so the actual trench must be deeper by the thickness of the wiring method chosen. Where *service conductors* are installed 18 inches or more underground, a plastic ribbon is required to be installed 12 inches above the conductors to provide a warning against digging into the conductors in the future.

Protecting with conduit at ground level   Wherever underground cable emerges from the ground, the above-ground portion must be specially protected against physical damage. Run the cable through conduit starting below grade. Install a conduit bushing at the bottom to protect the cable against damage from possible sharp edges where the conduit was cut. Where the cable enters the conduit, provide a short vertical S-curve in each wire to help prevent damage as the earth settles or moves under the action of frost. In addition, be sure to place an expansion fitting at the end or make other arrangements to accommodate the effects of ground movement on the conduit riser. As illustrated in Fig. 17–9, *NEC* 300.5(J) requires this to prevent riser movement through frost heaving or ground settlement from pushing through concentric knockouts or doing other damage.

The top of the conduit may end at the meter socket, in which case the conduit is terminated in the bottom of the socket using regular fittings. If it ends at a different location outdoors, install a service head of the type shown in Fig. 8–21. If it ends inside a building, terminate the conduit at a junction box as shown in Fig. 17–16, and continue with Type NMC cable from that point on. Of course, the conduit may end at the service equipment cabinet. If the conduit does not terminate in a grounded box or cabinet, the conduit must be grounded by a clamp or grounding bushing. If the conduit ends inside a building at a point no higher than the outside end of the conduit, fill the outside end with a sealing compound to prevent water from entering the building through the raceway.

INSTALLING SERVICE EQUIPMENT AT BUILDINGS

At the house and each building served by wires from the yard pole, there must be a service entrance similar to that described in Chapter 8, except that there is no meter to install (because the meter is at the pole).

Entrance at house   Let the service entrance cable or the conduit end at a point above the insulators; from there it runs directly to the service equipment inside the house. Where there is an underground metal water pipe, run the ground wire to it from the neutral bus in the service disconnect enclosure, and supplement the water pipe electrode with an additional electrode, probably a driven rod or pipe, bonding the two together. If there is less than 10 feet of metal water pipe underground, then the driven rod becomes the primary electrode, but the installation will look much the same, because the water piping *inside* the building must be bonded to the electrode and the neutral.

The wiring inside the house is the same as in other houses, and has already been covered in other chapters. Be sure to install plenty of circuits to allow for future expansion and to accommodate new appliances in the future. Spare circuits installed at the start are a good investment; adding them later can be costly.

Entrance at other buildings   How to determine the service entrance size for individual buildings is discussed earlier in this chapter beginning on page 190. If there are more than two circuits, the *NEC* requires a 60-amp switch as a minimum. Be sure to observe *NEC* grounding requirements as outlined earlier in this chapter. Note that per *NEC* 225.36 *Exception*, in garages and outbuildings on *residential* property, the disconnecting means may be an ordinary switch (single-pole, three-way, or four-way) as used in buildings serving primarily a residential purpose, as distinguished from other buildings used primarily in the business of farming. Buildings serving primarily a residential purpose might include a garage or smokehouse.

All other buildings must have a conventional disconnecting means that will open all the ungrounded wires. It may be located in or on the building, and may be either a switch or a circuit breaker, and it *must* be suitable for use as service equipment as evidenced by listing as SUITABLE FOR USE AS SERVICE EQUIPMENT. If the building is served by two 120-volt wires including a grounded wire, use a single-pole switch or breaker. If served by three wires (or only the two hot wires) use a double-pole switch or breaker. How to determine the ampere rating of the switch or breaker is covered later in this chapter.

Tapping service wires at buildings   If a building contains a substantial load, it should be served by wires direct from the pole. When two buildings are quite close to each other with neither building having a substantial load, both can be served by a single set of wires from the pole. Naturally the wires must be big enough for the combined load of both buildings. At the service insulators of the first building, make a tap and run the wires on to the second building, as shown in Fig. 17–6. At the second building, proceed as if the wires came directly from the pole.

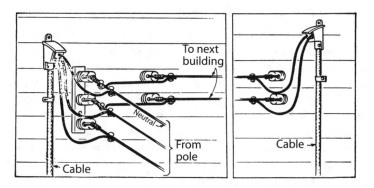

Fig. 17–6 When two buildings are near each other, tap the service wires from the pole where they are anchored on the first building and run them to the second building.

If the second building is very small and requires only 120 volts, tap off only two wires including the grounded wire, as shown in Fig. 17-7. If the second building has a considerable load so that 120/240 volts are desirable, tap off all three wires. The wires to the next building must be of the same size as the wires from the pole. A separate service switch and ground are required at the second building.

GROUNDING ELECTRODES—LOCATION, INSTALLATION, AND SPECIAL REQUIREMENTS

Fig. 17-7 Use a copper ground clamp with a copper rod.

Before proceeding with this topic, first review the information on grounding in Chapter 7 and on pages 91-96 in Chapter 8.

Grounding electrodes, where to install and how to wire Grounding electrodes must be installed at every building supplied by service conductors, including each building supplied directly from a distribution point. If one building supplies another building, install a grounding electrode at the second building unless the supply consists of just a single (either two-wire or multiwire) branch circuit.

At each building supplied by a service disconnect (except as noted in the next paragraph) interconnect the grounded conductor and the equipment grounding conductor and service enclosure at one point only, in the service equipment. Service equipment comes with the means to make this connection. Then connect the grounding electrode conductor so it runs directly from this point to the grounding electrode. This procedure could also be used at a building supplied from another building, but not for new construction and only if no common grounding return path exists between the two buildings, such as a shared metallic water service.

At all other buildings, connect the grounding electrode conductor to the equipment grounding conductor in the enclosure for the main disconnecting means for the building, leaving the grounded circuit conductor insulated from the equipment grounding conductor. This procedure must also be used when the service conductors include a separate equipment grounding conductor (See page 188).

Installing ground rods If other qualified electrodes are unavailable (See page 92), a ground rod can be used as a substitute. It will be usually at least 5/8-inch in diameter (1/2-inch if listed) with a steel core and a copper layer on the outside, driven at least 8 feet into the ground. Galvanized steel pipe is acceptable according to the *NEC*, but not always to the local inspector. Pipe must be at least trade size 3/4, and driven 8 feet into the ground. If the first electrode fails to achieve an earth resistance at least as low as 25 ohms (usually the case), an additional one must be driven at least 6 ft away. No further electrodes are required after the second is installed, regardless of the final resistance to earth. Connect the two rods together using wire the same size as the ground wire.

Bonding to water pipe and other electrodes   Note that if there is less than 10 feet of metal underground water pipe in direct contact with the earth, it does not qualify as a grounding electrode. Since metal water piping inside the building must be bonded to the grounding electrode, in practice you must in every case bond together the water pipe, regardless of length, and another acceptable electrode (grounded metal building frame, concrete-encased electrode, driven rod or pipe, or buried plate).

If the ground rod is copper-coated, use a clamp made of copper or brass; if the ground is galvanized iron, use a clamp made of galvanized iron. An iron clamp is shown in Fig. 8–23; one copper type is shown in Fig. 17–7, but there are others similar to that shown in Fig. 8–23.

Bonding raceways to enclosed grounding (or grounded) conductors   All metallic raceways must be bonded to the equipment grounding conductor of the enclosed circuit, or in the case of service raceways, to the enclosed grounded circuit conductor. If the raceway is continuous to an enclosure, the raceway termination at the enclosure will do this automatically. However, the metallic portion may be discontinuous, as in the common case where a metal elbow is used in a nonmetallic conduit run to prevent the pulling line from sawing through the inner radius of the elbow in a hard pull. The *NEC* addresses this problem in several ways. Isolated metal elbows need not be bonded if they are buried, measured to all points on the elbow, not less than 18 inches. In addition, on the load side of the service equipment only, metal elbows installed for this purpose need not be bonded if they are embedded in at least 2 inches of concrete. Finally, for isolated metallic portions of a riser on a pole, you can install a clamp on the metallic portion and extend a bonding conductor along the outside of the raceway up the pole as far as necessary to reach an accessible portion of the grounded or grounding conductor, as the case may be. Normally such bonding conductors cannot exceed 6 feet in length, but this is a special exception.

Burying the ground rod   Grounds must be permanent, so use extreme care in their installation. The *NEC* requires that at least 8 feet of a driven rod be in contact with earth, so a standard 8-foot rod and the connection to it must be buried. If you want the connection to be accessible, use a longer rod and build a box or provide a shallow well around the clamp location to protect it from physical damage. The ground rod is driven about 2 feet from the pole (or building) after a trench about a foot deep from rod to pole has been dug. The top of the rod is a few inches above the bottom of the trench. The ground wire runs down the side of the pole (or building) to the bottom of the trench, then to the ground clamp on the rod. After inspection, the trench is filled in and the rod, the clamp, and the bottom end of the ground wire remain buried. See Fig. 17–2. Remember that you will have a much better ground if you install two or three interconnected rods at least 6 feet apart.

Grounds in buildings housing livestock should be installed so that seepage from animal waste does not saturate the ground around the rod. Chemical action in time eats up the wire, the clamp, and sometimes even the rod, so what was once a good ground turns out to be no ground at all.

Equipotential planes—protecting against stray voltage   Due to the sensitivity of livestock to very small "tingle" voltages, the *NEC* requires an equipotential plane in livestock (does not include poultry) confinement areas if they contain metallic equipment accessible to animals and are likely to become energized. These areas must include wire mesh or other conductive elements embedded in (or placed under) a concrete floor, and those elements must be bonded to metal structures and fixed electrical equipment that may become energized, as well as the grounding electrode system in the building using 8 AWG (minimum) copper wire. This rule applies to all indoor locations, and outdoors in areas where concrete slabs support metallic equipment that may become energized and accessible to livestock, with the plane extending at least as far as required to encompass the area the livestock will be standing over.

Remember that the grounding system to which equipotential planes should be connected is usually (refer to the distribution point discussion earlier in the chapter) electrically separated from neutral return currents. The idea is to minimize voltage gradients by surrounding the livestock with bonded, conductive materials. To the extent stray currents enter the area, the conductive materials should collect them, equalizing the voltage across the area affecting any particular animal. Stray currents need not arise on the farm itself; there are well-documented examples of stray voltages entering farm (or other) environments from problems on local power supplier distribution systems. Remember, it doesn't matter if the voltage to true ground is 2.5 volts at the cow's rear hooves, as long as it is the same voltage at the cow's front hooves, the stanchion, and on the milking machine, etc. As long as that is the case, the cow sees zero volts and continues to produce milk. Due to the well-grounded environment, the *NEC* also requires all 15- and 20-amp, 125-volt, general-purpose receptacles in the area of an equipotential plane to have GFCI protection. This GFCI protection requirement also applies to similar receptacles in all damp or wet locations, including outdoors. The rule applies to "125-volt, single-phase 15- and 20-ampere *general purpose* [italics supplied] receptacles" and therefore would not include specific loads, such as a single receptacle on an individual branch circuit that is used exclusively as the supply for an electric fence voltage generator, or other comparable applications.

Surge protective devices   While lightning-caused damage to electrical installations is quite rare in large cities, it is rather frequent in rural areas. The more isolated the location, the greater the likelihood of damage. It occurs frequently on farms, and to a lesser extent in suburban areas and smaller towns. It is more frequent in southern areas, especially in Florida and other Gulf states.

Lightning does not have to strike the wires directly; a stroke *near* the wires can induce very high voltages in the wires, damaging appliances and other equipment as well as the wiring. Sometimes the damage is not apparent immediately, but shows up later as mysterious breakdowns. While proper grounding greatly reduces the likelihood of damage, a surge arrester correctly installed reduces the probability of damage to a very low level. Three leads come out of it; connect the white wire to the

Fig. 17-8 A surge (lightning) arrester installed in a service switch cabinet. Be sure its white wire is connected to the grounded neutral busbar in the cabinet. Take care to terminate its wiring at appropriate terminals; most main circuit breaker lugs are not listed for the combination of the small wires from the arrester and the much larger feeder or service conductors. Consult with the panelboard manufacturer for availability of appropriate terminations for the surge arrester. *(Square D)*

grounded neutral busbar, the other two to the hot wires. One should be installed at the meter pole. If the feeders from pole to building are quite long, install another at the building. Review the coverage in this book at the end of Chapter 7, then install a properly rated device in the service switch or panelboard, as shown in Fig. 17–8.

For service equipment using circuit breakers—and where there are two adjacent spare spaces for plug-in breakers—there is a surge arrester available that is shaped like a two-pole circuit breaker. It can simply be plugged into the main bus, with a wire tail to connect to the neutral bus. This removes the dilemma of where to connect the wires from the arrester described in the previous paragraph, because *NEC* 110.14(A) requires that terminals for more than one conductor be identified as such.

WIRING BRANCH CIRCUITS AND OUTLETS IN FARM BUILDINGS

The basic wiring principles covered in previous chapters apply to farm installations as well. In addition, your wiring design must take into account the special conditions that exist in farm buildings, particularly in buildings that house animals.

Number of branch circuits in farm buildings   How many circuits should a farm building have? Install enough circuits so that the business of farming can be carried on efficiently. Skimping leads to inefficiency and higher cost. Far better to install a panelboard with space for a few extra circuit breakers (or fuses) than to skimp now and pay a much higher cost per circuit later on. If a building is to be provided mostly with lights and a few incidental receptacle outlets for miscellaneous purposes, one or two circuits may be sufficient. A machinery shed may require only one circuit; a well-equipped farm workshop may require four; a dairy barn with water heaters, milking machines, milk coolers, and related equipment may require a dozen circuits and large service-entrance wires in proportion to the load. Be sure to leave provisions for additional circuits to be added in the future.

Cable for barn wiring   Type NMC nonmetallic-sheathed, corrosion-resistant cable, shown in Fig. 4–5, was developed for use in buildings with damp and highly corrosive conditions resulting from housing livestock. Each conductor has its own moisture-resistant insulation, and the several conductors are embedded in a plastic jacket that is specially resistant to moisture, mildew or other fungi, and corrosion. Type NMC may not be buried in the ground, but may be used anywhere that Type NM is acceptable. In the farmhouse, Type NM cable or any other wiring system may be used.

In some localities Type NMC is hard to find. In that case use Type UF cable described in this chapter. Type UF costs more than Type NMC, but may be used anywhere that Types NM and NMC may be used, and may also be buried directly in the ground. You can also use Type SE cable if it is made up with copper conductors. The more usual aluminum type cannot be used. Regardless of the wiring method selected, a separate copper equipment grounding conductor must be installed or be a constituent of the wiring method, and, if run underground, the wire must be insulated or covered. For cabled wiring methods, always secure them within 8 inches of the boxes or cabinets in which they terminate.

Raceways   For raceway methods, the usual method is polyvinyl chloride (PVC) conduit, covered in Chapter 12. If you need a flexible wiring method, liquidtight flexible nonmetallic conduit can be used.

Be sure to install expansion fittings in any locations where the air will not be conditioned, as in unheated buildings and outdoors. PVC as typically constituted usually expands roughly four to five times as much as concrete or wood or steel (including RMC) or most other building surfaces. One of the leading PVC conduit makers recommends allowing 140°F as a design temperature differential when working in areas with direct sunlight exposure, because sunlight heats the conduit more than the ambient temperature would predict. Referring to *NEC* Table 352.44(A), this would cause almost 6 inches of possible movement over a 100-foot run. How much of that appears at any given installation depends on the time of year it was first installed. For example, if it was installed in a moderate temperature on a cloudy day, it probably was in the middle of its size range, and might be expected to contract 3 inches during the coldest night and expand 3 inches during the hottest afternoon.

For PVC the movement resulting from ordinary outdoor temperature changes cannot be safely accommodated with conventional fittings over even comparatively short distances. Such movement is enough to break out concentric knockouts, break enclosures free of their supports, and do other damage. The *NEC* insists on expansion fittings for this material if you expect it to exceed more than ¼ inch in total expansion/contraction between fixed points—that amount of movement would occur over less than a single 10-foot conduit length in most climates. Plan on installing these fittings routinely on all outdoor or barn installations unless the basic arrangement accommodates the movement in other ways. For example, for a straight RNC service riser (assuming the service head doesn't butt against the roof), no expansion fitting would be required. However, the appropriate mounting hardware must be used so this movement can occur safely. This movement is also beyond the anticipated effects of ground movement, covered in "Protecting with conduit at ground level" earlier in this chapter, although one properly installed expansion fitting could usually cover both requirements. Figure 17–9 illustrates these points.

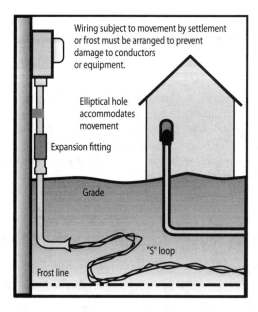

Wiring subject to movement by settlement or frost must be arranged to prevent damage to conductors or equipment.

Elliptical hole accommodates movement

Expansion fitting

Grade

"S" loop

Frost line

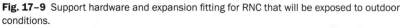

Fig. 17-9 Support hardware and expansion fitting for RNC that will be exposed to outdoor conditions.

Selecting boxes, plates, and sockets   Metal outlet boxes, plates or covers, and lighting sockets carry a risk of shock, particularly in the damp conditions frequently encountered on farms. Nonmetallic wiring methods and boxes do not suffer the corrosion problems common to metallic systems, and as a consequence they are far more common on farms. *NEC* 547.5(A) generally requires such wiring methods in agricultural buildings (or metal-clad cable with a nonmetallic jacket over the armor).

In the installation of nonmetallic boxes there is a precaution to observe. Wood swells and shrinks in locations where dampness and humidity levels vary. Steel boxes, if solidly mounted on supporting timbers, do not present a mechanical problem as the timbers swell with moisture, because the steel can give a little if required. But nonmetallic boxes, if screwed down tightly on dry timbers, have been known to break out their bottoms as timbers swelled with increasing moisture. So if you mount nonmetallic boxes on the surface of dry timbers, leave just a little slack; don't drive the mounting screws down completely tight.

Outdoor switches and receptacles   Ordinary receptacles and switches are designed for indoor use. If installed outdoors without further protection they would be very unsafe. But they can be used outdoors if installed in special waterproof boxes of cast iron or aluminum with threaded hubs instead of knockouts. Outdoor receptacles must have weatherproof covers; and, unless they are of a locking configuration, they must observe the weather resistance rating requirement covered on page 117.

Boxes for outdoor switches and receptacles   Boxes for installing outdoor receptacles are made with a variety of threaded-hub arrangements instead of knockouts. The boxes have letter designations for the different numbers and locations of the hubs. Examples are Type FS with one hub at the top and Type FSC with a hub at the top and another at the bottom of the box. Ordinary 15-amp or 20-amp toggle switches or GFCI-protected receptacles up to 50 amps may be installed in the boxes, using the proper weatherproof (gasketed) cover. If using Type UF cable outdoors, use a weatherproof connector similar to that shown in Fig. 17–10, and be sure the cable is marked SUNLIGHT RESISTANT.

Covers for outdoor receptacles   The *NEC* in 406.8 specifies two types of weatherproof receptacle covers for outdoor (and other wet location) receptacles, the choice depending on how the receptacle is to be used:

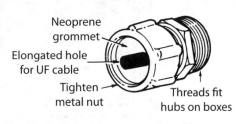

Neoprene grommet

Elongated hole for UF cable

Tighten metal nut

Threads fit hubs on boxes

Fig. 17–10 Watertight connector for cable entry into threaded hub.

■ For loads that are attended while being used (such as portable tools), and which exceed 20 amps or 250 volts, the enclosure may be the type with flip-covers that are weatherproof only when the lids are closed. An example of this type is shown in Fig. 17–11. This type may also be used for any receptacle in a damp location, meaning located where protected from beating rain, such as under a roofed open porch or canopy.

■ For loads that are not attended when they are being used, or for any receptacle in a wet location rated 15 or 20 amps at 125 or 250 volts, the enclosure must be weather resistant whether or not a plug is inserted. An example of a "while in use" cover is shown in Fig. 17–12. These covers will be the usual ones required for these applications. Note that they usually are not rated to prevent a hose stream from entering, as in the case of milking parlors. For such applications, the *NEC* allows a cover that is watertight only when the plug is removed.

◀ **Fig. 17–11** Flip-type receptacle covers that are weatherproof only when lids are closed. *(Leviton Manufacturing Co., Inc.)*

Fig. 17–12 Hooded receptacle cover that is weather resistant with or without plug inserted. *(TayMac Corporation)* ▶

Poultry, livestock, and fish confinement systems   These areas, a subset of agricultural buildings generally, have been plagued by accumulations of litter and feed dusts, which are often corrosive and likely to infiltrate into electrical components. These areas require dustproof and weatherproof enclosures for this reason, along with the restricted list of wiring methods. In addition to these concerns, poultry houses have specialized lighting needs.

Protecting against dust, moisture, and vapors   Accumulation of dust, particularly dust with moisture, inside ordinary boxes has been identified as the cause of many fires in buildings where poultry, cattle, and hogs are housed. *NEC* Article 547, Agricultural Buildings, specifies requirements related to this safety issue. Consult your local electrical inspector when there is any question whether a particular building is subject to these requirements. The *NEC* says that where animals are housed, special precautions must be taken when any of the following is present: dust from litter or feed; water from cleaning operations or condensation; corrosive vapors from animal waste; or any combination of these.

Use molded plastic boxes as shown on the left in Fig. 9–18, with close-fitting gasketed covers. Type NMC or UF cables must be supported within 8 inches of boxes, and must have water- and dust-tight entries. Use a threaded compression watertight connector similar to that shown in Fig. 17–10, making sure the opening in the neoprene grommet is the right size for your cable. Rigid polyvinyl-chloride (Type PVC) conduit is well-suited to the conditions of dust, moisture, and corrosive vapors. Rigid metal conduit, intermediate metal conduit, or EMT may be used as the wiring method where only dust and/or moisture is present, but are not recommended where corrosive vapors are present.

Motors must be totally enclosed or designed to minimize the entrance of dust, moisture, or corrosive particles. Lighting fixtures must be guarded where exposed to physical damage; they must be designed to minimize the entrance of dust and moisture; and if exposed to water they must be watertight—see Fig. 17–15.

Poultry house lighting   Special wiring is required for lighting designed to promote egg production. It is well known that hens produce more eggs during the winter if light is provided before sunrise and after sunset. Opinions vary as to the ideal length of the "day," but a 14-hour or 15-hour period seems reasonable, which means that the length of time the lights must be on varies from season to season and must be adjusted about every two weeks.

If all the lights are turned off suddenly in the evening, the hens can't or won't go to roost, but will stay where they are. It is necessary to change from bright lights to dim lights to dark. This can be done by manually operated switches, but an automatic time switch costs so little that manual switches should not be considered. The wiring for such switches is simple and wiring diagrams are furnished with the product.

Be sure that both the bright light and the dim light fall on the roosts, because if the roosts remain in darkness when the lights come on, the hens may not leave their roosts. Neither will they be able to find the roosts in the evening if the roosts are in darkness. One 40-watt or 60-watt lamp with a reflector for every 150 to 200 square feet of floor area is usually considered sufficient for the "bright" period. Normal spacing is about 10 feet apart. For the "dim" period, 10-watt lamps are suitable.

Barn wiring   Barns are generally very humid, especially in winter. Proper ventilation will greatly reduce this humidity, but few barns are sufficiently ventilated. The moisture will naturally collect in the coldest parts of the barn, and in winter that means the outside walls. If possible, avoid running cables on outside walls where the alternate wetting and drying might damage the wiring. However, if you use Type NMC or Type UF cable with nonmetallic boxes, the likelihood of damage is almost entirely eliminated.

It is best not to run cable along the bottoms of joists or other timbers because it might be subject to mechanical injury. Don't run it at right angles across the bottoms of joists even if running boards are used. The cable will receive far more protection if you run it along the side of a beam, then along the side of a joist to the middle of the aisle to each point where a light is to be installed, as shown in Fig. 17–13.

It will take less cable (and lead to less voltage drop) if you run cable down the middle of the aisle through bored holes in the joists; the bored holes should not be near the extreme edge of the joist. The exact method will depend on the details of carpentry in the barn being wired. It is important to install cable so that it is protected as much as possible from damage and from excessive moisture.

Receptacle and lighting outlets   Receptacle outlets should be installed where they will not readily be bumped by animals. Most barns have too few outlets. The right number will depend largely on the kind of barn being wired. Install enough receptacle outlets so that extension cords need not be used.

Do not skimp on the number of lighting outlets to be installed. The preferable number is one behind each animal stall. The minimum is one behind each pair of stalls. Do not install lighting outlets on the bottom of timbers, but preferably between joists so that the bottom of the lamp is flush with the bottom of the joist. Damage to the lamp is less likely to occur that way. Lights should be controlled by toggle

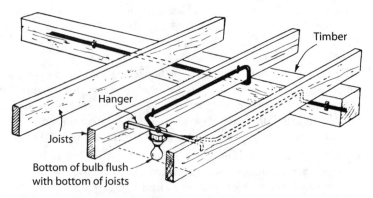

Fig. 17–13 Run cable along the substantial timber to prevent later damage. The bottom of the lamp should not project beyond the bottom of the timber.

switches located in protected spots and as high as practical so they cannot easily be damaged by animals. If you mount switches at elbow height, you can operate them even if your hands are full. For convenience, have at least some of the light controlled by three-way or four-way switches located at various entrances.

Reflectors maximize light output   When exposed lamps are used, half of the light goes downward, and the other half goes upward and strikes the ceiling. Ceilings in barns and similar buildings are usually dirty, so most of the light directed at them is absorbed rather than reflected. Use a good reflector for each lamp to redirect light downward. A 60-watt lamp with a good clean reflector usually gives as much useful light as a 100-watt lamp without the reflector. Reflectors are inexpensive; one type is shown in Fig. 17–14. Clean the reflectors regularly.

Haymow fixtures and wiring   Good illumination of the stairway or ladder to the haymow will help prevent accidents. Haymow lights should be controlled by a switch on the main floor. For convenience, install a pilot light at the switch.

The dust in a haymow is combustible and a fire can occur if an exposed lamp is broken. When a lamp breaks it burns out, but during that fraction of a second while it is burning out the filament is at an exceedingly high temperature (4,000°F) and can

Fig. 17–14   Reflectors pay dividends. A 60-watt lamp with a reflector gives as much useful light as a 100-watt without a reflector. Keep reflectors clean.

start a fire. For haymows it is best to use gasketed enclosures known as "vaporproof" fixtures, as shown in Fig. 17–15. Glass globes enclose the lamps. In many localities use of vaporproof fixtures or "dust-ignition-proof" fixtures is required in haymows.

The *NEC* requires that nonmetallic-sheathed cable must be given special protection when it runs through a floor. In a haymow, there is always danger that a pitchfork will puncture unprotected cable and cause a short circuit. It is reasonable that inspectors require extra protection from the floor up to a point where hay will never be in contact with the cable. Occasionally an inspector will permit cable that is installed in the corner formed by wall and stud with a piece of board nailed over it for protection.

More often, the cable is run through a piece of pipe or conduit. Note that this does not constitute "conduit wiring." The cable is pulled through the conduit and continues to be cable wiring, with the conduit merely furnishing mechanical protection for the cable. But you must ream the ends of the conduit to remove any burrs that might have been formed, and then install a

Fig. 17–15 Use protected fixture in haymow wiring.

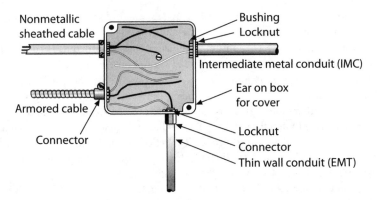

Fig. 17-16 How to change from one wiring method to another. Install a cover on the box upon completion.

bushing on each end.

In some localities it is the custom to wire the entire haymow with rigid conduit or EMT; the cable then ends on the main floor. This is a safety measure protecting the wiring in the haymow against mechanical injury that might easily occur if cable is installed where exposed to hay forks. The conduit is not as likely to rust out in the haymow where it is not exposed to fumes and moisture as on the main floor. When wiring with conduit, the change-over is easily made as shown in Fig. 17–16.

NEC 250.86 requires that you ground isolated sections of conduit in most cases, unless it is "short" (250.86 Exception No. 2) and used only for cable support and protection. That is easily done by connecting the bare grounding wire of the cable to a green grounding screw in one of the holes in the outlet box where the conduit begins.

FARM SAFETY

In rural areas it is wise to take precautions against the hazards of darkness, lightning, and fire. Consider the following safety additions for your installation.

Installing yard lights Every farm will have one or more yard lights for convenience and accident prevention. Figure 17–17 shows an efficient and long-lasting high pressure sodium yard light, photocell-controlled to turn on automatically at dusk and off at dawn. It can be mounted on a pole or high up on a building wall. Unless automatically controlled, a yard light should be controllable from at least two points: from the house and from a barn or other building. This requires three wires as shown in Fig. 10–18. It is contrary to the *NEC* to feed a

Fig. 17-17 Photo-cell controlled, high pressure sodium yard lights contribute to convenience and safety on the farm.

yard light by tapping the wires on the meter pole. In addition, being made of metal, it must be grounded to an appropriate equipment grounding conductor.

A properly installed yard light requires a considerable quantity of wire and other materials, especially if it is controlled from several points. Ordinary 14 AWG wire may be electrically large enough, but often a larger wire must be used for mechanical strength for long overhead spans. It is practical to install low-voltage, remote-control switching as discussed on page 216. With respect to relays, if you provide photocell control, be sure the switching system complies with *NEC* 410.104(B) and opens all ungrounded conductors to the lamp ballast. If the photocell operates only as a single-pole device (most do), it can operate a relay but it cannot be used to control one of these fixtures connected line to line. This is not an issue on 120-volt circuits, but will be on a 240-volt circuit.

Bonding a lightning protection system   Some farms employ a lightning protective system that consists of lightning rods (officially termed "strike termination devices") on the roof and heavy conductors running down to a driven rod electrode. The grounding electrodes for both the lightning protective system and the electric service must be: 1) different electrodes, but then 2) those different electrodes must be all bonded together. Installing lightning protection systems involves a different NFPA standard and requires specialized training.

Water pump—fire safety considerations   Every farm will have a water pump. It serves to provide water for all the usual purposes and is a tremendous help in case of fire. But during a fire, quite often power lines between buildings fail and fuses blow so that the pump cannot run just when it is needed most. That failure can be avoided by adding an independent feeder to the pump location, as if it were a fire pump. Simply arrange another feeder from the distribution point directly to the pump, preferably using an underground wiring method for reliability. Fire pumps are exhaustively covered in *NEC* Article 695, and given the fact that this pump would be routinely used for other purposes, it could not for many reasons technically qualify as such. Nevertheless, most inspectors would say that the second feeder to the pump location is permissible under *NEC* 225.30(A)(1). You would, however, need to put a sign at both the pump disconnect location and the main building feeder disconnect location explaining at each location that an additional source of local supply is located at (specify the other location and its function). That way no one will be deceived into thinking the entire building has been disconnected after opening switches at only one of the two locations.

Chapter 18
LOW-VOLTAGE WIRING

LOW-VOLTAGE SYSTEMS DISCUSSED HERE OPERATE at less than 100 volts, with most at well under 100 volts. Telephones and doorbells are common examples. Others include CATV, lighting outlets using transformers, wireless radio-controlled switches, central vacuum system controls, some thermostats, and data processing signaling circuits. The technically correct term is limited-energy or limited-power systems, because even a 6-volt system with enough amperage running through small wires is a fire hazard. But low voltage is a more familiar term to many.

Critical safety considerations apply to the systems in this chapter. First, make sure the power source to any of these systems, even a doorbell transformer, is listed as "Class 2," or is derived from listed computer equipment and used for data interchange, such as computer network wiring. This means it has been investigated for safety from fire or shock hazard. The only exception is communications circuits (telephone systems) connected to a central station (or wired in a similar manner), but all network-connected telecommunications equipment must be listed.

Second, never place Class 2 (or telephone or computer network) wiring inside a power raceway, even if you use one of the wire types suitable for line voltage. Class 2 wiring may enter an enclosure to terminate at a relay that controls a power circuit, but even then it must be restrained so at least a ¼-inch air separation is maintained from the power wiring. Normally, however, listed equipment has two entirely separate compartments, or else a solid grounded barrier between the two classes of use. For example, if you are wiring a furnace, run the power circuit in one of the wiring methods described in Chapter 12. Run the Class 2 control circuit separately. For convenience, *NEC* 300.11(B)(2) allows a power raceway to support, on its outside, a Class 2 cable if that cable is used to control the equipment supplied by the power circuits in the raceway. This permission does not apply to any other raceways, and never to power cable assemblies.

TELEPHONE WIRING
The separation point between the telephone company's network and the customer's

premises is called the demarcation point. Everything on the network side of this point is the property and responsibility of the telephone company. Equipment and wiring on the customer side are the customer's responsibility. In some places the telephone company installs the demarcation point equipment, consisting of connection and testing terminals, required grounding, and equipment for protection against lightning and voltage surges. In other cases it is done by the user. Check with your telephone company about local practice. You can have the telephone company do the wiring inside your home, but the charges are typically high. Instead, you could hire an electrical contractor or a communications contractor to do the work inside your home, or you can do it yourself.

Safety precautions   If you decide to install telephone wiring yourself, several safety precautions should be observed.

- Make sure the telephone ground is to the same electrode as the power ground. If it isn't, make sure that the two electrodes are bonded together with a 6 AWG copper wire.

- Disconnect the telephone line at the demarcation point or take a telephone handset off the hook to prevent the phone from ringing. Telephone voltages can reach dangerous levels when the phone is ringing.

- Do not install or use a wired telephone during a lightning storm because of the danger of shock to yourself and damage to equipment.

Telephone and CATV cable   Listed cables include Types CMP (telephone) or Type CATVP (Community Antenna Television) for use in air-handling ducts or plenums; CMR (or CATVR) for risers between floors; CM and CMG (or CATV) for general use but not for plenums or risers; and CMX (or CATVX) for use only in dwellings (also permitted if installed in raceways).

When stripping telephone wire, keep in mind that the 22 to 24 AWG sizes are relatively delicate so care is needed to avoid nicking the wire. It is safer to use terminals that displace the insulation rather than ones that require it to be stripped.

Wiring methods for multiple telephone outlets   There are two general methods of wiring for several telephone outlets. One is to run everything in parallel (as are the lights in Fig. 6–7). Troubleshooting at a later date on such a system is difficult. The other method is to run a separate cable to each telephone outlet from the demarcation point. This method makes it easier to upgrade later.

Copper conductors are twisted together in pairs to minimize induction and crosstalk and to reduce interference from nearby sources such as motors. Two, three, or more pairs of conductors are run in one cable. Although only one pair is needed to operate a single telephone, running multiple pairs to each outlet will accommodate future uses such as intercom, speed dialing, call forwarding, multiple lines, and other features at one telephone. One pair could be used for a computer modem, another for a fax machine. The standard color code for twisted pairs is green/red, black/yellow, and white/blue. There is increasing use of more sophisticated station wire that uses color banding with complementary colors in each twisted pair. For example, the red/green pair is replaced by a blue wire with white bands and a white wire with blue bands; the black/yellow is replaced by orange with white and white with orange; and the white/blue is replaced by green/white and white/green. When making connections, always match the colors of the wires to the color markings on the terminals of outlets and on other telephone equipment.

New construction   Consider adding two or more telephone lines in new construction so you can have separate lines for uses such as computer modems and fax machines. Also install plenty of telephone outlets whether they appear to be needed or not. For convenience, every floor including the basement should have telephone outlets. These outlets can consist of pull wires from near the demarcation point to each outlet. The outlet needs only to be a plaster ring covered with a blank plate that can later be changed to a telephone jack when needed. The *NEC* requires a minimum of one wired communications outlet within every dwelling unit.

The telephone jacks that you install at the wall outlet are larger than the jacks built into the telephone bodies and handsets. Wall jacks are simple to connect. Each component will come with clear installation instructions.

Alternatively, cable can be installed if the pull-wire method is impractical. The best way to do the job is to use outlet boxes and empty electrical metallic tubing or electrical nonmetallic tubing which allows for changing the cable at some future time.

Place telephone cable at least 12 inches away from the electrical cable running through your house to minimize electrical noise. Don't mistake electrical cable for telephone cable, which is about the size of a drinking straw and is very flexible. The telephone cables should cross power cables at 90 degrees. Do not have telephone wiring share the same holes through wood members or the same stud spaces with electrical wiring.

Old work   To meet the needs of a growing family or a home office, you can install additional jacks and lines. From the demarcation point, telephone cable can be run on the surface using wall jacks for surface mounting. You may find that running it along the baseboard and around door and window openings is sufficiently inconspicuous. Place the cable so it is not subject to physical damage. For a neat installation, install outlet boxes for telephone jacks and run the cable behind the walls or woodwork as described in Chapter 13, "Modernizing Old Wiring." The outlet box is not required for a telephone jack, but it provides a base for fastening the faceplate. There are also telephone jack faceplates that provide a means of attaching the jack to the plate without using a box.

GROUNDING

Both telephone and CATV systems share a likelihood of contact with induced voltages from lightning and contact with high-voltage power lines. Therefore most telephone systems must include a primary protector, and CATV systems a shield grounding terminal, as near as practicable from the entry point. A grounding connection must be extended from these devices to the power system grounding electrode system over the shortest and straightest practicable path. In the case of one-

and two-family dwellings, the length must not exceed 20 feet unless an additional electrode is installed within the 20-foot limit, and that electrode is then bonded to the power system electrode using a 6 AWG or larger copper conductor.

HOME COMPUTER NETWORK WIRING

Most new homeowners anticipate being able to connect multiple computers together to take advantage of high-speed Internet access and share data access to certain electronic appliances. For example, providing a high quality laser printer at every computer location may be prohibitive, but the same objective can easily be met if every computer can connect to one such printer at a central location. Commercial operations do this routinely.

The solution is to connect the computers together in a network, ideally by hard wiring the building with cable that will accommodate this duty. There are wireless modes of data transmission, but they are not as fast as a permanently wired connection and are inherently less secure. The standard approach is to wire for "Ethernet" connections between computers. This approach relies on a four-pair cable, with each pair tightly twisted together. The terminating jacks and patch cables to the individual workstations are readily available at office supply and electronics stores. These jacks require "punch down" tools to install the wires into insulation-displacing terminals, because the required degree of twists cannot be maintained otherwise. If you are making only a few terminations the plastic tools available with the jacks are adequate, but if you are doing a whole building, buy the specialized tool that pushes the wire into the terminal and trims its end in one operation.

The degree of performance for data transmission over cables and equipment is described by categories. Ordinary voice calls ("plain old telephone service" or "POTS" in the trade) do fine on Category 1. Ordinary computer modems need a higher standard, and high-speed data transmission requires the comparatively finicky Category 5, 5e (for extended), or even 6, requiring a very high standard of workmanship to maintain successfully. Even the slightest lapse can make the difference between a successful 100 MB Ethernet connection or only a 10 MB connection. Make sure your cable and all terminals meet the required standard. Alternate methods promising greater bandwidth probably represent the way of the future, such as coaxial cable service ("cable modems"), and optical fiber using laser-generated light transmitted over special glass or other fibers.

DOORBELLS, CHIMES, AND OTHER LOW-VOLTAGE CONTROLS

These necessary items are easy to install. The power is furnished by a transformer, a device that reduces 120-volt ac to a different ac voltage. For doorbells this is usually about 6 to 10 volts. Doorbell transformers have two "primary" *wire* leads which are permanently connected to any point on any 120-volt circuit that is not controlled by a switch, and two "secondary" *screw terminals* for the low voltage. The power consumed while the bell is not ringing is under one watt, despite the permanent connection to the line. When you push the button to ring the bell, you close the low-voltage circuit, and then the transformer draws about five watts from the 120-volt line.

Installation of doorbells and chimes is discussed here. For instructions on repairing a doorbell or chime that does not ring, see pages 227–228 in Chapter 19, "Troubleshooting and Repairs." The same type of wiring can be used for low-voltage thermostats for air-conditioning and heating systems. There are many different control systems in use; follow the wiring diagrams provided with the equipment.

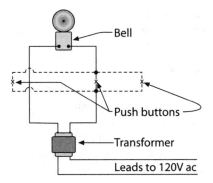

Fig. 18–1 The wiring of a doorbell with a transformer is simple.

Doorbells   For installation of a doorbell, consider the two secondary or low-voltage terminals on the transformer as the SOURCE for a circuit, the bell as the outlet, and the button as a switch. Breaking it down this way makes all these diagrams very simple. Fig. 18–1 shows the installation of one bell with a push button in one location. If a button is wanted in an additional location, install it as shown in the dotted line, in parallel with the first button.

In most installations, a single bell is not considered enough. One choice is to install a bell for the front door and a buzzer for the back door. You will see that Fig. 18–2 is the same as Fig. 18–1 for a single bell, with the buzzer now added as shown with the dotted lines.

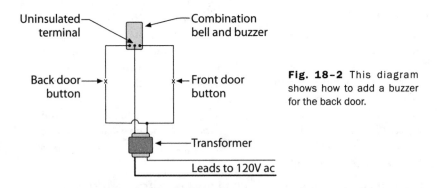

Fig. 18–2 This diagram shows how to add a buzzer for the back door.

Door chimes   Musical chimes are often used instead of doorbells. Models are available that sound two musical notes when the front door button is pushed, and a single note when the back door button is pushed. No matter how long the button is held down, the sound does not repeat. The wiring is the same as for the doorbell.

If you install chimes in place of an existing doorbell and find that the sound is not very loud, you will have to install a new transformer. The transformers used with doorbells usually deliver only about 6 to 10 volts, and many chimes require from 16 to 24 volts.

Because the usual transformer used for doorbells and chimes cannot deliver more than about 20 volts, nor more than about 5 watts, there is little danger of shock or fire even in case of accidental short circuit. Therefore the wire used for this low-voltage wiring does not need much insulation. Bell wire, which is insulated with only a thin layer of plastic, is commonly used. It does, however, need to be listed for suitability in terms of not adding to the potential fuel load in a fire. See Fig. 18–3. Attach it to the surface using insulated staples.

Fig. 18–3 Shown above are wires for connecting doorbells and chimes. This cable is listed as "CL2" which means it can be used for most general-purpose applications, but not in risers and plenum cavities in highrise or commercial occupancies, which are beyond the scope of this book.

Wireless radio-controlled doorbell system   An alternative to the wired-in doorbell or chimes is a radio-controlled doorbell system. The sounding unit is usually electronic, and some can be programmed to play different tunes. There will be a battery, probably at the pushbutton location, which will need replacing periodically. A convenient advantage of the radio-doorbell is that the sounding unit can be taken with you to the basement, patio, or backyard so you can hear the bell wherever you are.

LOW-VOLTAGE SWITCHES

Using a transformer provides an extremely flexible system for control of lighting outlets from any number of switch locations. The 120-volt branch-circuit wires are run to each lighting outlet, but all the switch legs are low-voltage wires—easy and economical to install.

In each outlet box where power is consumed, a very small relay (which is an electrically operated switch) is installed with one end projecting out of the outlet box through a knockout. A 24-volt transformer installed in some convenient location furnishes power to operate all the relays. Because the voltage is low and the transformer delivers only a very few watts even if short-circuited, the wiring between the transformer, the relays, and the individual switches resembles doorbell wiring. Inexpensive wire, insulated for only a low voltage, is run from the transformer to each relay, then to as many switches as you wish. The wire does not need to run through conduit nor does it have to be cable of the armored or nonmetallic-sheathed type. Run it or fish it through walls and staple as required. The switches do not need outlet boxes. Each switch is really a double pushbutton. Pushing the top end of the button on any switch turns on the light; pushing the bottom end turns it off.

Fig. 18-4 The low-voltage (small) end of this latching relay projects out of a ½-inch knockout in the outlet box on which the fixture is mounted. Some models are solid-state (electronic) with no moving parts. *(General Electric Co.)*

Fig. 18-5 The type of switch used in remote-control wiring is shown above. The same kind of switch is used at every location. It is equivalent to two push buttons. Use as many switches as you wish to control one outlet (or several outlets if you wish). *(General Electric Co.)*

Fig. 18–4 shows one of the relays, Fig. 18–5 one of the switches, and Fig. 18–6 the general type of wire used. Fig. 18–7 shows the wiring circuit.

You have a choice of procedures for installing the relays. One way is to run the low-voltage wires to each box, letting at least 6 inches project into the box through the knockout in which the relay is to be later installed. The second approach is to install the relays in the boxes first, and then connect the low-voltage wires to them later, allowing at least 6 inches slack in those wires. Unless the extra length of wire is provided, you would later find it impossible to replace any relay that proves defective. As an alternate to installing relays at the controlled outlets, relays can be ganged in a centrally located box which has separate compartments for the circuit

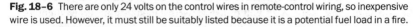

Fig. 18-6 There are only 24 volts on the control wires in remote-control wiring, so inexpensive wire is used. However, it must still be suitably listed because it is a potential fuel load in a fire.

wiring and the low-voltage wiring.

WIRELESS RADIO-CONTROLLED SWITCHES

If you have a hallway, stairway, basement, or attic where it would be convenient to have the light controlled by three-way switches, but the addition of the second switch and the wiring to it seems like too big a job, consider a radio-controlled set of three-way switches. One switch—the receiver—replaces the present single-pole switch. The other switch—the sender—is operated by a battery. It is surface-mounted, so

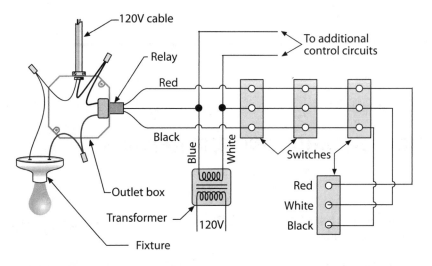

Fig. 18-7 Typical remote-control circuit, in this case with the outlet controlled at four locations. One transformer is used for the entire house. The same circuit is often used for remote control yard lights.

no cutting into the walls is required. These wireless switches operate just as standard three-way switches do—the light can be turned on or off from either location—but they are much easier to install.

Switching of loads plugged into receptacles can also be done by radio-operated switches. The receiver plugs into a wall outlet, and the load plugs into the receiver. The sender can be up to 50 feet away. If you use more than one of these radio-controlled switches, be sure each operates on a different frequency so they will not interfere with each other.

CENTRAL VACUUM SYSTEMS

The motor of a central vacuum system is started by the closing of a low-voltage control circuit when an inlet is opened and the vacuum hose connected. A metal tank containing the motor and a collection bag is permanently wired in a utility room or in an attached garage. Airtight hose about 1½ inches in diameter is run to wall inlets in convenient locations. The wall inlets are gasketed so that when not in use they do not leak air into the system. Newer units sometimes also have a 120-volt connection at the hose inlet, designed to accommodate a line-voltage beater bar at the attachment end. These require a hard-wired connection to a local lighting or receptacle circuit. Remember, however, that the 20-amp small-appliance circuits (Refer to this topic in Chapter 5) must not have any other loads connected to them, which means if one of these outlets is in the kitchen or dining room, the power for the hose attachment must not come from the adjacent receptacle circuit.

Chapter 19
TROUBLESHOOTING AND REPAIRS

THIS CHAPTER EXPLAINS HOW TO diagnose problems with existing wiring and how to make common repairs.

FOR PROBLEMS WITH—	CHECK TOPIC HEADING—
■ Aluminum wiring	*Updating aluminum wiring*
■ Chimes	*Troubleshooting doorbells and chimes*
■ Circuit Breakers	*Troubleshooting fuses and circuit breakers*
■ Cords	*Troubleshooting cords and plugs*
■ Doorbells	*Troubleshooting doorbells and chimes*
■ Fuses	*Troubleshooting fuses and circuit breakers*
■ Lamps—floor and table	*Repairing table and floor lamps*
■ Light fixtures—ceiling: fluorescent and incandescent	*Troubleshooting light fixtures*
■ Plugs	*Troubleshooting cords and plugs*
■ Receptacles—wall: two-wire and three-wire	*Testing and replacing wall receptacles*
■ Switches—wall: single-pole and three-way	*Replacing wall switches*

TROUBLESHOOTING FUSES AND CIRCUIT BREAKERS

The different types of fuses and circuit breakers are reviewed in Chapter 5 on pages 44–46. A fuse blows or a breaker trips for one of two reasons. Either something connected to the circuit is defective, thus drawing an excessive number of amperes, or there are too many lights, appliances, or motors connected at the same time, overloading the circuit.

If a fuse blows or breaker trips quickly every time a particular appliance is plugged in, especially if it makes no difference whether it is plugged into a different circuit, the appliance is defective. Often the defect is in the cord. If the appliance has a removable cord, try a different cord; if the cord is permanently attached, only careful inspection will locate the defect. Badly twisted and worn cords must be replaced, not repaired. See "Troubleshooting Cords and Plugs" in this chapter on page 225.

If a fuse blows when a motor (as in a home workshop) is turned on, remember that a motor that consumes only 6 amps while running may draw over 30 amps for a few seconds while starting. Substituting a time-delay fuse for the ordinary fuse might solve the problem. If the fuse continues to blow, suspect the motor. Check the cord. Check to see that the motor bearings have oil. (Some motors have sealed bearings never requiring oil.) Perhaps the belt is too tight, or the machine that the motor drives lacks oil, increasing the load on the motor and increasing the amperes beyond the safe point.

If fuses blow or circuit breakers trip from time to time on any one circuit for no apparent reason, it is likely that the circuit is simply overloaded. You will need to disconnect some of the load on that circuit. The wise procedure would be to install an additional circuit. Alternatively, the trouble may be in the wiring of the circuit protected by that particular fuse or breaker, and the short circuit or ground must be located and repaired.

Diagnosing blown fuses   Usually by looking at a fuse you can tell if it is blown and what made it blow—an overload or a short circuit.

Plug fuses   To tell if a plug fuse is blown, look through the glass window on top. In a normal fuse you will see a small link of metal, usually with a narrower portion in the middle, as shown in *A* of Fig. 19–1. If the fuse has blown because of an ordinary overload, it will have the appearance of *B* in the same illustration; the narrow part of the link is gone. If the inside of the window is spattered with metal as in *C*, it has blown because of a short circuit or ground fault. In that case, you must investigate and correct the cause before replacing the fuse.

Time-delay fuses   If your fuses are of the time-delay type shown in Fig. 5–3, note the coiled spring in a normal fuse. All Type S nontamperable fuses have this feature, as do some conventional plug fuses shown in Fig. 5–2. If such a fuse blows, look first at the view window and then, if possible, at the spring. If the viewing window is obscured because of spattered metal, the fuse has blown because of a short circuit or ground fault. If the window is relatively clean, look at the spring; if the spring is contracted so that its coils touch each other, it has blown because of an overload.

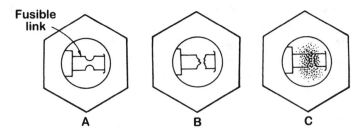

Fusible link

A   B   C

Fig. 19–1 Plug fuse: *A*—normal; *B*—blown because of overload; *C*—blown due to short circuit.

Cartridge fuses These rarely show any outward sign of having blown. Test in place with a neon tester, or remove the fuse from the fuseholder and test it with a continuity tester.

The *neon tester* in Fig. 19–2 can be used to test fuses in their holders if you can get at the terminals of the fuseholder. With the circuit on, touch the two leads of the tester to the two terminals of the fuseholder. If the light in the tester glows, the fuse is blown. If it does not glow, the fuse is not blown. If the cartridge fuses are installed in a fuse-holding pullout block, note that the pullout block usually has small holes through which the leads of the tester can be inserted without removing the block from its holder.

Fig. 19–2 This inexpensive neon test light can be used on either 120-volt or 240-volt circuits and is also handy for testing fuses. It contains a tiny neon bulb that consumes only the smallest fraction of a watt and lasts many thousands of hours. *(A. W. Sperry Instruments Inc.)*

A *continuity tester* is pictured in Fig. 11–7. To use this type of tester you must first remove the fuse from the fuseholder using a plastic fuse puller. If the fuses are in a pullout block they can safely be removed from the pullout block without the need for a fuse puller. Then touch the tester's prod to the metal ferrule or blade at one end of the fuse, and touch the alligator clip to the other end. If the lamp lights, the fuse is good. If the lamp does not light, the fuse is blown.

Resetting circuit breakers Most breakers are reset after tripping as shown in Fig. 5–6. A few brands are made so the handle returns to the off position when the breaker trips, and it is only necessary to move it back to the on position (instead of forcing it beyond off before returning it to on as in most brands).

Labeling fuses and circuit breakers The *NEC* requires (in 408.4 and 110.22) just the sort of sort of circuit directory presented here for safety reasons.

Specifically, circuit directories must show "the clear, evident, and specific purpose or use" of each of the circuits, and the description must include sufficient detail so that "each circuit [can be] distinguished from all others." The circuit directory must designate spare overcurrent devices as just that because the absence of a label would create confusion. Finally, circuit directories must not refer to "transient conditions of occupancy" in their descriptive language. Most circuit directories long outlive such terminology as "John's bedroom" or "Bev's Boutique." Use points of the compass, diagrams, handedness, or other permanent methods to describe connected loads. A circuit directory might read something like this:

 1 Service disconnect
 2 Large bedroom and hall lights and receptacles
 3 Small bedroom lights and receptacles, and bathroom lights
 4 Bathroom receptacles
 5 Receptacles in living room and ceiling lights in kitchen and dining room
 6 Bottom halves of all kitchen receptacles, and all receptacles in dining room
 7 Top halves of all kitchen receptacles
 8 All basement lights and receptacles
 9 Range
10 Clothes dryer
11 Water heater
12 Clothes washer
13,14 Spares for future new circuits

UPDATING ALUMINUM WIRING

"Old technology" (pre-1972) aluminum wire in the branch circuit sizes, 10 and 12 AWG, had a coefficient of expansion that caused it to expand and flow away from beneath a terminal screw when heated by current running through it. Then, when it cooled, the connection would be loose. To compound matters, oxygen would get to the aluminum at the loose terminal, and aluminum oxide, a non-conductor, would form on the exposed aluminum wire. Not every termination behaved in this way, but enough did that by November 1971 the aluminum alloy used in electrical conductors was changed and the terminations on switches and receptacles were redesigned. Some aluminum wiring installations may need to be updated in order to prevent the possible development of loose connections, excessive overheating, and fire. If the 15-amp and 20-amp branch circuits in your dwelling were wired with aluminum conductors, a review of the section on aluminum wiring on page 33 will be helpful as you prepare to evaluate your installation. Inspect all of your receptacles and switches to see if they are marked CO/ALR. If they are not, you should replace them with devices having the CO/ALR marking, or reconnect them with copper pigtails.

In replacing the switches and receptacles, if the aluminum wire is copper-clad any kind of terminal may be used. (Terminals are discussed on pages 37–40.) Push-in terminals, shown in Fig.4–16, may be used with 14 AWG copper or copper-clad aluminum, but *not* with ordinary aluminum. Receptacles marked CO/ALR are not easy to find and may have to be special ordered.

The alternative method is to "pigtail" a short copper jumper to connect to each switch or receptacle terminal, *being sure to use splicing devices UL listed for use with both aluminum and copper in combination.* This way the aluminum circuit wire will not depend on the device terminals for continuity. A pigtail splice is shown in Fig. 4–15, but the twist-on splicing device shown there may not be UL listed to connect copper and aluminum. One that is listed is a tool-applied pressure connector that requires a trained electrician to do the work. This is not a do-it-yourself job for the average homeowner. There are listed twist-on wire connectors that make this a much simpler proposition than in the past. Be sure to follow the installation instructions carefully, particularly regarding the allowable wire combinations. For example, at this writing all the available wire combinations available in the listed device include at least one copper wire along with the aluminum ones.

Consult a qualified electrical contractor if you feel your aluminum-wired house needs attention and you cannot find CO/ALR devices or the new connectors suitable for use with aluminum wires.

REPLACING WALL SWITCHES

Wall switches can wear out with constant use. Replacement is simple. First turn off the main switch or breaker for the building, or at least unscrew the fuse or trip the breaker that protects the circuit on which the switch is installed. Never work with hot wires. Remove the faceplate over the switch, then remove the two screws holding the switch to the box. Pull the switch out of the box.

Replacing single-pole switches   If the switch you are replacing is a single-pole type, it will have only two terminal screws. Loosen the screws, remove the wires, and connect them to the two screws of your new switch. It makes no difference which wire goes to which terminal. Be sure the loop on your wires is turned clockwise around the screw (See Fig. 4–13). Reinstall the switch in the box and reinstall the faceplate.

Replacing three-way switches   If the switch you are replacing is the three-way type, it will have three terminal screws. One of them will be a dark, oxidized color, much darker than the others. In disconnecting the wires, make careful note which wire runs to the dark-colored screw of the old switch; that wire must run to the dark screw on your new switch. It makes no difference to which screws the other two wires are connected.

Some switches do not have terminal screws; instead the wires are pushed into openings on the switch. To remove them, just push a screwdriver with a small blade into the slot near the openings for the wires and pull them out.

TESTING AND REPLACING WALL RECEPTACLES

To determine if a receptacle is live, insert the two leads on the neon tester shown in Fig. 19–2 into the two openings of the receptacle; if the lamp lights, the circuit is live. If the lamp doesn't light, you should consider whether the receptacle could be worn out. Its interior contacts may be bent so they no longer make good contact with the blades of a plug, in which case the receptacle should be replaced. If your receptacles are of the grounding type, as in Fig. 7–11, insert one lead of the tester into the round opening on the receptacle, and the other lead first into one of the two parallel openings, then the other. If the receptacle is properly installed, the tester should light when the second lead of the tester is inserted into the narrow slot that leads to the hot wire to the receptacle, but not when it is inserted into the wide slot connected to the grounded wire.

As in replacing a switch, turn off the main switch or breaker of the building, or at least unscrew the fuse or turn off the breaker that protects the circuit on which the receptacle is installed. Remove the faceplate, remove the two screws holding the receptacle to the box, and pull the receptacle out of the box. Note that it will have two silver-colored terminal screws (called the "white" terminals) and two brass terminal screws. Like switches, some receptacles have no terminal screws. Remove the wires as described under switches. If your new receptacle has terminal screws, connect the white wires to the silver-colored screws, the others to the brass screws.

If the receptacle location is one that requires GFCI protection under the present *National Electrical Code* (*NEC*)—bathroom basins; outdoors; at kitchen counters or bar, laundry, or utility sinks; garages; in crawl spaces and attics having equipment requiring servicing—the replacement must be GFCI-protected.

Replacing two-wire receptacles   If the receptacle you are replacing is the old-fashioned kind (for two-prong plugs only), the replacement method depends on the type of wire used in your house. If your house uses wiring methods that provide for equipment grounding, replace the receptacle with a grounding type receptacle. Run a wire from the new receptacle's green terminal to the box in which it is installed, unless the receptacle is of the type specially designed to be acceptable for use without the grounding wire. If the box is not grounded, you may replace the receptacle with another non-grounding type, or with a three-wire receptacle grounded to the grounding electrode system or to the grounding electrode conductor. You may also use a GFCI receptacle, but you must mark it (or the wall plate) no equipment ground. You may also use a conventional grounding (three-wire) receptacle if there is GFCI protection ahead of it, either as a feed-through GFCI device or as a GFCI circuit breaker. In this case, mark the receptacle location no equipment ground and GFCI protected.

In either case, note which wire(s) run to the brass terminal screws of the old receptacle, and connect them to the brass screws on the new receptacle. Connect the other wire(s) (which should be white) to the silver-colored screws, fold the wires behind the receptacle (See "Preventing loose connections" on page 172), and reinstall in the box.

Replacing three-wire receptacles   If your receptacle is the grounding type for three-prong plugs, be sure the same wire that connects to the green screw of your old receptacle is connected to the green screw of your new receptacle. *Unless this is done, the receptacle is wrongly connected and can be very dangerous.* If there is no bond wire, make sure that the box is grounded, and then either use a receptacle with a self-grounding yoke or make the bonding connection as shown in Fig. 11–5.

Don't forget that for most receptacles in a house, when providing a conventional three-wire receptacle replacement, the rules in the *NEC* for tamper resistance (See page 223) apply to the replacement receptacle, and for outdoor locations in any occupancy the requirements for weather resistance apply as well. Similarly, any receptacle, whether or not three-wire, replaced in a location where current *NEC* rules mandate GFCI protection must have that protection as well, either through a GFCI circuit breaker or by using a GFCI receptacle.

Testing for faults in three-wire (i.e., with ungrounded, grounded, and grounding wires) circuits A three-wire circuit analyzer, shown in Fig. 19–3, will indicate if the three-wire circuit is correctly wired. Possible faults—open ground, open neutral, open hot, hot/ground reverse, and hot/neutral reverse—are indicated by various combinations of lights on most testers. Codes on the analyzer package will give the reason for the wiring fault so you will know how to fix the problem. The reversed connections are easily corrected. The opens may be difficult to track down, but it is very rare that a conductor in a cable is open between outlets. The open will probably be found at a poor splice in a junction box or a poor connection at a receptacle. From the outlet where the tester shows an open, go back halfway to the panelboard and test again. If the open still shows up, you know the trouble is

Fig. 19–3 A three-wire circuit analyzer is used to test for faults in the wiring to three-wire circuit receptacles. *(A. W. Sperry Instruments, Inc.)*

between that point and the panelboard. By dividing the circuit run in half and testing the remainder each time, you should be able to locate the trouble.

TROUBLESHOOTING LIGHT FIXTURES

Be sure to observe any maximum lamp size marking on fixtures, and relamp with the proper size. This is particularly important with recessed fixtures and with pan-type fixtures having lamps close to the ceiling. Many fires have started due to the heat from too-large lamps with too-high wattage.

Repairing fluorescent fixtures The usual fixture for residential use is 4 feet long with two 34-watt tubular lamps. Some very old fixtures may have a starter for each lamp. The starter will be found near the end of and behind each lamp and can be rotated out and replaced easily. If replacing the starter does not eliminate the trouble, or if the lamps are blackened at the ends, replace the lamps. The usual bi-pin lamp is removed by rotating the lamp 90 degrees in either direction and then sliding the lamp out of the socket. Replace with the same length and wattage as marked on the old lamp.

A lamp socket that is cracked or broken must be replaced. Be sure the circuit is disconnected before working on a socket. The pan or cover over the fixture body must be removed to get at the sockets and their wiring. Carefully observe how the socket is supported and connected. Take the old one to the store to be sure of a match, and install the new socket.

If it still does not light, and you have eliminated everything but the ballast (See pages 15–18 for a discussion of the operation of fluorescent lights), you might consider replacing the whole fixture. Replacement ballasts often cost more than a new fixture. To replace a ballast, first disconnect all the wires and then loosen the supports, usually two bolts and wing nuts. Again, be sure the replacement matches both electrically and physically. Make sure the fixture is grounded properly. The new fixtures designed for compact fluorescent lamps roughly the size of incandescent lamps also have ballasts, but they are much smaller than the traditional ones for long tubes. The *NEC* requires a disconnecting means (for other than dwellings) so ballasts can be safely changed with the circuit on, usually applied as a small, inexpensive, pull-apart unit in the ballast channel. This rule applies, and the disconnects must be installed when existing ballasts are changed.

Some old fluorescent desk lamps have a pushbutton switch that functions as a starter. The button must be held down for a few seconds to pre-heat the lamp. Other than this switch, the other elements are similar to those of the larger fixtures.

Repairing incandescent ceiling fixtures   If your incandescent ceiling fixture does not light, the first thing to do is to check the lamp in the fixture, and replace it if it has burned out. The next suspect is the wall switch, which may have failed. Replacing the wall switch is covered earlier in this chapter. It is possible the center contact in the lamp socket may have lost its spring due to heat, or it may have become corroded. With the power off, gently lift the center contact to restore some of the spring in it, and clean off any corrosion with a bit of sandpaper. Loose connections to the circuit wires are a possibility but not a common occurrence—an improperly installed solderless connector may have come loose.

TROUBLESHOOTING CORDS AND PLUGS

Never yank on a cord to unplug something because this puts all the strain on the fine wires inside the cord. Sooner or later some of the strands of the wire will break, and you will have a short circuit. A fuse will blow and your cord will be ruined. Instead, grasp the cord body itself and pull it out of the receptacle. Unplug appliances with removable cords at the receptacle rather than at the appliance. When removing the cord from the appliance, grasp the cord body itself instead of pulling on the cord.

Damaged cords   There is a serious danger of both fire and shock from a worn or damaged electrical cord. Be sure to do an occasional safety check of your extension cords, appliance cords, and cords on floor and table lamps, heating pads, etc. Signs of cord damage include a frayed or worn outer jacket or the appearance of damage to the wires inside the jacket. Using such a cord might be compared to driving a car with thoroughly worn-out tires that might blow out at any time—it is an unwise and unsafe policy. Don't try to repair the cord—replace it instead if possible, as on a lamp. For items like a heating pad where the cord cannot be replaced, discard the entire item and replace it with a new one.

Damaged plugs   If a cord is unusable only because the plug on its end has been damaged but the cord otherwise appears sound, it can be repaired by installing a new plug if the shortened cord is still adequate for its purpose. If a plug's blades are loose, bent or corroded, or if the body of the plug is cracked, replace it. Always replace a two-wire plug with a two-wire, and a three-wire with a three-wire, noting carefully that the green wire goes under the green screw, the black wire under the brass screw, and the white wire under the silver-colored screw. If your replacement two-wire plug is polarized (one blade narrower than the other) and your cord is Type SPT-2 (See Fig. 4–7), the wide blade and the grounded conductor must be connected together. But in Type SPT cord, where all of the insulation is the same color, it is difficult to know which is the grounded conductor. The grounded conductor is identified by either a raised ridge or a groove found in the insulation along one of the wires.

Be sure the replacement plug has a "dead front" with no live parts exposed after it is wired. This is the only type currently listed by testing labs. The old style with a separate fiber disk should not be reused. It has been discontinued because the "dead front" design is much safer.

The finely stranded wires used in flexible cords require considerable care in making terminations. Twist these strands together tightly before making a termination to prevent a loose strand from crossing where it does not belong.

Some means must be provided to prevent stress on the cord from being transferred to the terminals in the plug. This can be a cord-grip built into the plug (more likely to be found in heavy-duty plugs), a knot in the cord, or tape on the cord that will prevent a pull on the cord from being transmitted to the terminations.

REPAIRING TABLE AND FLOOR LAMPS

When the light doesn't come on, first eliminate a burned-out lamp bulb as the cause by testing with one you know is good. If the light still does not come on, the next most common failure is the switch that is built into the socket. The switch can be a push-through, a turn-knob, or a pull-chain type. Failures might also result from broken wires in the cord, bad connections, or dry and brittle insulation at the socket. Sometimes the contact, which is found at the center of the socket, becomes corroded.

Socket and switch problems   To inspect the socket and wires, unplug the portable lamp and remove the shade and the harp (wire shade-holder). The socket is made up of two parts that must be separated. The socket includes the threaded lampholder and switch, and an outer brass shell that is lined with a fiber insulating sleeve. They fit into a cap that remains with the portable lamp. To separate the shell from the cap, press where the shell is marked press and use an easy rocking motion to take them apart.

If the fiber insulating sleeve inside the brass shell has become dry and brittle, it should be replaced. If it is necessary to replace the sleeve or the entire socket, take the old ones with you to the store to be sure of getting the right part.

If the center contact on the socket is corroded, the corrosion can be removed by using fine sandpaper. Some of the original "spring" of the center contact can be restored by gently lifting it up with a small screwdriver.

A defective switch will be revealed by the test procedure in the following paragraph.

Cord and wire problems   Make a visual inspection of the cord. If there are worn spots in the insulation, or places where the cord has been tightly kinked, replace the cord. To check for broken wires in the cord use a continuity tester (See Fig. 11-7). Test between the center contact in the lamp socket to the narrow blade on the plug, and from the threaded shell of the lampholder to the wide blade of the plug. In both cases, the test light should go on unless the switch is turned off, in which case the test to the center contact will show open (no circuit). Turn the switch and try again. If it still tests open, either the switch is faulty or there is a poor connection at the socket or the plug. Test for a broken wire in the cord by making these same tests but between the plug blades and the wires in the cord, leaving the socket and switch out of the circuit.

Also inspect the insulation of the wires of the cord at the socket. If the insulation appears dry and brittle, this section of cord should not be used and needs to be replaced. If the cord is long enough (in some portable lamps, especially those with two or more sockets, there are splices at the top of the stem or in the base), it may be possible to feed through a couple of inches and cut off the bad wire. Use the remaining good wire now at the socket to make connections to the terminal screws. If the cord is too short to pull through enough good wire, replace the cord altogether. Unless the route from the lamp base to the lamp socket is a straight shot, it would be wise to use the old cord to pull in the new. Strip about 1½ inches of insulation from all four wire ends, form staggered hooks on the wires, and tape over them taking care that the resulting splice is not too bulky. Feed in the new cord at one end and pull out the old cord at the other.

Reassembly   After performing the above steps that may be necessary to repair your lamp, the socket/brass shell must be reassembled to the cap. Be sure you hear a "click" when the two parts go together. Then replace the lamp bulb, the harp, and the shade.

Broken lamp bulb   Sometimes a lamp bulb breaks, leaving the lamp base in the socket with nothing much to grip to unscrew it. To remove the base, be sure the portable lamp is unplugged or the circuit is off. Wear gloves and safety glasses. Using a cork that will just fit inside the base of the broken lamp bulb, push the cork hard into what is left of the base, and it will grab the remains so you can unscrew the lamp base from the socket by turning the cork.

TROUBLESHOOTING DOORBELLS AND CHIMES

Installation of doorbells and chimes is discussed on pages pages 213-215. To diagnose problems, begin by checking the various components—bell, pushbutton, transformer, and wiring—until you find the part that does not work and needs to be repaired or replaced.

Testing doorbells   To check a doorbell, start at the source and check the transformer. It has two screw terminals delivering low voltage—6 to 8 volts—for the bell, which is too low to give you a shock. Try to find a meter or tester that will work on voltage this low, and test for voltage. Only if you cannot find a tester, short-circuit these terminals momentarily using two short pieces of wire. Connect them to the terminals, strip the other ends, and then snap one of the wires across the second wire so contact is made and broken in a small fraction of a second. If you see a small spark, the transformer is not defective. This spark may be very faint, so conduct this test in the dark. If the wires remain shorted for more than an instant, you will burn open any overcurrent protection built into the transformer by its manufacturer as part of its Class 2 listing, ruining it because this protection is not replaceable by design. If you get no spark, check to see that the 120-volt circuit supplying the transformer primary is alive. If it is, the transformer has burned out and must be replaced. See Figs. 18–1 and 18–2 and related text. Be aware that this test can easily ruin a perfectly good transformer; always try to get a meter or tester instead.

If the transformer is all right, the next most likely source of trouble is the pushbutton. Remove the mounting screws of the pushbutton and disconnect it from the wires. Touch the ends of the two wires together. If the bell rings, the button is at fault. If you can reach the contact surfaces, clean them with fine sandpaper and adjust them so they make good contact but will open when the button is released. Reconnect the pushbutton and try it. If the bell still does not ring, replace the pushbutton.

If both the pushbutton and the transformer are all right, check the bell or buzzer itself. Disconnect and remove the bell or buzzer and take it to the transformer location. With a couple of short wires, connect the bell or buzzer directly to the transformer. If it rings, then the trouble is in the wiring. If it does not ring, the bell or buzzer is defective and must be replaced.

Chimes—transformer and plungers   If you have replaced a bell with chimes, it may be that the transformer was not changed. Chimes require a higher voltage— 18 volts or more—than a bell or buzzer. Install a new transformer. If the chimes still do not work, check the plunger(s) for free movement. Over time, plungers get gummed up with oily dust and are not free to move when the pushbutton is pushed. Use a cotton swab and solvent to clean up the plunger(s), being careful to keep the solvent off the solenoid coil wires. Remember that when testing chimes, they will operate only as contact is made—they do not ring continuously as a bell does, but "ding" once each time the pushbutton is depressed.

Wiring between components   If the pushbutton, transformer, and bell (buzzer, chimes) all check out, then the trouble is in the wiring between them. These wires are small and subject to breakage when flexed, so check at the ends where they have been worked over during installation to see whether there is a broken wire. Due to the low voltage, the insulation on these wires is adequate electrically but not very resistant to physical damage. Look for locations where the insulation has been damaged, which you can repair with a small piece of electrician's tape. Much of the wiring is probably concealed in the walls, and if the damage is concealed there is little that can be done short of abandoning the old wires and fishing new wires in as replacements. That may be the time to look into installing a wireless, radio-controlled doorbell system described on page 215.

About the Authors

H. P. RICHTER wrote the first edition of *Wiring Simplified* in 1932. It was one of the first how-to books on wiring for general readers. It was followed by *Practical Electrical Wiring* in 1939. With the publication of each succeeding edition of these books, the author came to be regarded more and more widely as both the do-it-yourselfer's electrician and the electrician's electrician. He had the rare ability to write for both the amateur and the professional, and his books continue today in updated versions.

W. CREIGHTON SCHWAN, registered professional engineer and electrical consultant, was a recognized authority on wiring standards and practices. He was the coauthor of *Wiring Simplified* and *Practical Electrical Wiring* for several editions and regularly contributed a column, "Code Comments," to the magazine *Electrical Contractor*. His background included positions supervising electrical inspectors in Alameda County, California, and acting as electrical safety engineer for the California Division of Industrial Safety and senior field engineer for nine western states for the National Electrical Manufacturers Association. He held memberships in several professional organizations and had been an instructor at seminars on the *National Electrical Code*.

CURRENT AUTHOR FREDERIC P. HARTWELL is nationally recognized for his editorial and technical skills on *NEC* topics. He has been responsible for more than 1,000 successful proposals and public comments regarding changes in the *NEC* over the past nine *Code*-making cycles. He continues as a member of *National Electrical Code* Committee, serving as the senior member of Code-Making Panel 9. He is a licensed master electrician and the secretary of the Massachusetts Electrical Code Advisory Committee.

He is the former senior editor and *National Electrical Code* expert with *Electrical Construction and Maintenance Magazine*. He is the current author of *Practical Electrical Wiring* from Park Publishing. He has been the author of the *American Electricians' Handbook* published by McGraw-Hill (since the 15th edition published in 2008), and authored the 26th, 27th, and 28th editions of *McGraw-Hill's National Electrical Code Handbook* covering the 2008, 2011, and 2014 *National Electric Code*.

GLOSSARY

Alternating current (ac).   Current that periodically reverses, having alternately positive and negative values.

Ambient temperature.   Temperature of the medium (air, earth, water, etc.) surrounding conductors, devices, or utilization equipment.

Ampacity.   The maximum current, in amperes, that a conductor can carry continuously under the conditions of use without exceeding its temperature rating.

Ampere (A).   Unit used to express the flow of electricity; one coulomb (the measure of quantity of electric power) per second = one ampere. Water analogy: gallons per minute.

Approved.   Acceptable to the inspector.

Arc.   Unwanted flow of electricity through an insulating medium (such as air), characterized by the emission of light and heat (as opposed to controlled arcs, such as in electric-discharge lamps).

Arc-fault circuit interrupter (AFCI).   A device that recognizes circuit characteristics unique to arcing faults, and then opens the circuit path to prevent the fault from continuing. Arcs may occur between circuit wires, or across a discontinuity in one of the circuit wires, or between an ungrounded circuit wire and a grounded object.

Armored cable.   Two or more conductors, paper wrapped, enclosed in a flexible spiral interlocked armor of steel or aluminum, and with an aluminum bonding strip under and in contact with the armor. Informally referred to as "BX."

AWG (American Wire Gauge).   Standard for measurement of wire diameters in the United States.

Ballast.   Auxiliary equipment used with fluorescent and electric-discharge lamps to provide starting voltage and proper operating voltage, and to limit the current.

Bonded (Bonding).   Connected to establish electrical continuity and conductivity.

Branch circuit.   The wiring that connects an outlet or group of outlets to the last fuse or circuit breaker. (*See also* Feeder)

Building.   A structure that either stands alone, or that is cut off from all adjoining structures by firewalls and, if applicable, approved fire doors. A firewall is typically masonry or concrete and is expected to survive and remain standing after a conflagration on either side. It differs from a fire separation assembly, which need only prevent the spread of fire for a stipulated period. For example, the separation between adjacent dwelling units in "townhouse" style construction may or may not be actual fire walls, depending on the requirements of the local building code. Therefore adjacent dwelling units, or groups of dwelling units in such construction may or may not qualify as separate buildings. Only if so qualified are they eligible for separate services by right under *NEC* requirements.

Bus, busbar.   Conductor, usually a bar of rectangular cross section, serving as the common connection for two or more smaller conductors. Part of panelboards, including those used in service equipment.

Bushing.   Metallic or insulating fitting used at the end of a raceway to provide a smooth surface for wire leaving the raceway and entering an enclosure.

BX.   Trade name for Type AC armored cable.

Cable.   Assembly of two or more wires, as in nonmetallic-sheathed or metal-clad cable. Individual large-size wires are also referred to as cables because they are made up of several strands of smaller wire.

Capacitance.   Measure of the ability of a circuit component to store electric charge; exhibited by capacitors, two conductors separated by a nonconductor. Capacitance in an alternating current circuit makes the current run ahead of the voltage. (*See also* Reactance)

Circuit.   Arrangement of conductors, devices, and utilization equipment (loads) such that current will pass through them.

Circuit breaker.   Switching device, manually operable, which automatically opens a circuit at a predetermined level of overcurrent. (*See also* Fuse, Branch circuit)

Class 1 (or 2 or 3) control circuit.   (*See* Low voltage)

Codes.   Local, state, and national regulations governing safe wiring practices. (See also *National Electrical Code*)

Conductor.   Material, usually metal, having relatively low resistance, through which current will readily flow. Examples: wires, cables, busbars.

Conduit.   Steel, aluminum, brass, or nonmetallic raceway of circular cross section through which wires can be pulled. Protects wires from physical damage.

Current.   Flow of electricity. (*See also* Alternating current, Direct current, Leakage current)

Cycle.   In alternating current, the time (usually 1/60 second) during which the voltage goes from zero to + maximum, back through zero to maximum, and back to zero again. Measured as Hertz (Hz.)

Demand. Kilowatts consumed during a specified period, usually 15, 30, or 60 minutes.

Demand factor. Ratio of the maximum demand to the total connected load.

Device. Electrical system component designed to carry, but not consume, electrical energy. Examples: switch, receptacle, circuit breaker.

Direct current (dc). Current that flows in one direction only, characterized by "poles" termed positive and negative; generally supplied by batteries, but sometimes by generators.

Disconnecting means. Equipment used to de-energize (disconnect from the source of power) an entire electrical installation or individual circuits and appliances; may consist of circuit breaker or fused switch, or the plug and receptacle in the case of small portable appliances.

Duplex receptacle. Electrical apparatus consisting of a pair of receptacles on a single yoke for plugging in two different appliances.

Enclosure. General term for outlet box, junction box, or cabinet, whether metallic or nonmetallic, enclosing wires, devices and/or electrical equipment

Equipment grounding conductor. (*See* Grounding conductor, equipment)

Feeder. A circuit extending between the service equipment (or other power supply source) and the final branch-circuit overcurrent device. (*See also* Branch circuit)

Fish tape. Narrow, highly tempered steel tape with a loop at one end for pulling wire through conduit and pulling cable through walls and other spaces. Also available in nonmetallic configurations.

Fittings. The various standardized parts used in a wiring system to make mechanical connections. Examples: anti-short bushings, solderless connectors, conduit couplings.

Fixture. (*See* Luminaire)

Four-way switches. Switches having four terminals for wires and used with a pair of three-way switches to control a single light from more than two locations.

Fuse. A soft metal link that melts and opens a circuit at a predetermined level of overcurrent. Fuses are enclosed in a convenient case that contains the melted metal when a fuse blows, making replacement easy.

Gang. To join individual electrical devices or boxes into a single unit. Example: To mount two devices side by side, two single boxes can be converted to one "two gang" box by removing one side of each box and bolting the boxes together.

Greenfield. A trade name for flexible metal conduit available in both steel and lightweight aluminum. Also called "flex." Similar to the armor of armored cable but the wires are pulled in after the flex is installed.

Ground. The earth.

Ground fault.   Unwanted current path to ground.

Ground-fault circuit interrupter (GFCI).   Device intended for protecting people against shock by sensing abnormal current to ground and then interrupting the circuit. Usual trip threshold is 4–6 milliamperes.

Ground-fault protection for equipment (GFPE).   Device intended for protecting equipment by sensing abnormal current to ground and then interrupting the circuit. Operates above levels that would constitute shock protection.

Ground wire.   (*See* Grounding electrode conductor; grounding conductor, equipment)

Grounded (Grounding).   Connected (connecting) to earth or to a conductive body that extends the ground connection.

Grounded wire (identified conductor).   In a circuit, the wire (usually white, or gray, which is often used on 480Y/277 volt systems) that normally carries current and is intentionally connected to the ground at the service equipment, and is normally not fused or protected by a circuit breaker or interrupted by a switch.

Grounding conductor, equipment.   The conductive path(s) installed to connect normally noncurrent-carrying metal parts of equipment together and to the system grounded conductor or to the grounding electrode conductor, or both.

Grounding electrode conductor (ground wire).   A conductor used to connect the system grounded conductor or the equipment to a grounding electrode or to a point on the grounding electrode system.

Harmonics.   Multiples of the fundamental (60-Hz) frequency caused by nonlinear loads (e.g. 180-Hz, 300-Hz). Odd-order harmonic currents do not cancel in the neutral of three-phase, four-wire systems.

Hertz (Hz).   Unit used to express frequency. One hertz = one cycle per second. What used to be called "60 cycles per second" is now called "60 hertz."

Hickey.   (1) Metal coupling for supporting a lighting fixture from a fixture stud. The center is open to allow wires to enter the fixture stem. (2) Hand tool for bending rigid metal conduit.

Hot.   Designating any point in the electrical system having a voltage above ground potential. Every point should be considered to be hot until you have tested it for the presence of voltage.

Identified conductor.   (*See* Grounded wire)

Impedance.   The opposition to the flow of current in a conductor, device, or load in an alternating current circuit. It is measured in ohms, and can be represented as the hypotenuse of a right triangle whose horizontal leg is resistance and whose vertical leg is reactance. The power factor is the cosine of the angle between the resistance and the hypotenuse of the triangle thus formed. (*See also* Power factor)

Inductance.   Measure of the electromotive force (voltage) produced in a conductor through the action of a varying magnetic field produced by the varying currents associated with alternating current circuits. Inductance in an alternating current circuit causes the current to lag behind the voltage. (*See also* Reactance)

Insulation. Material having relatively high resistance, through which current will not readily flow when the material is applied within its voltage rating. Used to enclose the metal conductor of a wire.

Labeled. Product having an attached label, symbol, or other identifying mark indicating compliance with an appropriate standard administered by a testing organization.

Lamp. Produces visible light when an electric current passes through it. Includes incandescent (not "bulb"—that is the enclosing glass envelope), fluorescent, halogen, mercury and sodium vapor, and other types.

Leakage current. Small undesirable current that flows through insulation whenever voltage is present.

Listed. Included in a list published by a testing organization, indicating that the product meets appropriate standards.

Load. Anything that is connected to a branch circuit, feeder, or service and consumes electricity.

Low voltage. As used in this book, control and communications circuits with power supplies that strictly limit the amount of energy at specified voltages that a circuit can deliver. Generally, Class 2 control circuit limitations consider both shock and fire potential; Class 3 and communications circuits consider fire initiation potential. Class 1 control circuits may operate at line voltage and generally require conventional wiring methods.

Lumens. Measure of the amount of light produced by a lamp. One lumen falling on a surface of one square foot produces one footcandle.

Luminaire. The accepted international term for a complete lighting unit including the lamp and the associated parts required to position and protect the lamp (and ballast if applicable) and to connect the assembly to the power supply. This term is used in the *NEC* instead of the former "fixture" or "lighting fixture." Both terms are still used in this book.

Multiwire branch circuit. A branch circuit comprised of two or more ungrounded conductors (having a voltage between them) and a grounded conductor, with the voltage between the ungrounded conductors and the grounded conductor being equal.

National Electrical Code (NEC). Installation code for safe electrical wiring, sponsored by the National Fire Protection Association and revised on a three-year cycle. The 2014 *NEC* is the current edition.

Neutral. The wire, usually grounded, usually having white (or gray, especially common on 480Y/277-volt systems) insulation, related to two or more conductors of the circuit so that it has the same potential to each of them, and so that, with a balanced load and no harmonic currents present, it carries no current. It is connected to the neutral point of the system and, as of the 2008 *NEC*, it retains its status as a neutral even where it is only associated with a single ungrounded conductor.

Neutral point   The common point on a wye-connection in a polyphase system or midpoint on a single-phase, 3-wire system, or midpoint of a single-phase portion of a 3-phase delta system, or a midpoint of a 3-wire, direct current system.

Ohm.   The unit that is used to express electrical resistance or impedance. Ohms = volts ÷ amperes.

Outlet.   A point on the wiring system at which current is taken to supply utilization equipment. It is the limit of fixed wiring; the receptacle body at the end of an extension cord is not an outlet, but a permanently installed receptacle into which the extension cord is connected is an outlet.

Overcurrent.   Current in excess of the rated current of equipment or the ampacity of a conductor, which may result from overload, short circuit, or ground fault.

Overcurrent protective device.   Circuit breaker or fuses to protect conductors and equipment against overcurrent.

Overload.   Operation of equipment in excess of normal, full-load current rating or of a conductor in excess of rated ampacity for a sufficient length of time to cause damage or dangerous overheating.

Panelboard.   A panel including buses and overcurrent devices for the control of all branch circuits and sometimes feeders to other panelboards. In dwellings, the service disconnecting means and main overcurrent protection are often included in the branch circuit panelboard. Usually supplied with a suitable enclosure.

Parallel circuit.   Connection of two or more devices or loads across the same conductors of the circuit, current flow through each being independent of the others.

Pigtail.   Splicing a separate length of wire to two or more conductors to provide a tap point. (*See also* Splice, Tap)

Plug.   Male connector attached to a flexible cord for insertion into a receptacle; also called attachment plug and attachment plug cap.

Polarized plug.   Plug having one of the parallel blades wider than the other so it can be inserted into a polarized receptacle only one way, preventing accidental reversal of hot and grounded sides of circuit.

Power factor.   The ratio of true power (or watts) to apparent power (or volts × amperes), expressed as a percentage. Where the loads are inductive, such as motors and fluorescent lighting, the current lags the voltage resulting in a power factor of, typically, 80 to 90 percent. (*See also* Impedance)

Premises wiring.   The interior and exterior wiring that extends from the end point of the wiring under the control of the power supplier, or from any other source of supply such as a generator, to the outlets. It does not include internal wiring of appliances or comparable equipment.

Qualified person. A person with the skills and knowledge related to construction and operation of electrical equipment and installation, and who has received safety training to recognize and avoid the associated hazards.

Raceway. Wiring enclosure, metallic or nonmetallic, usually of a type having means for pulling conductors in and out. Conduit, electrical metallic tubing, electrical nonmetallic tubing, flexible conduits, surface raceways, and wireways are common types of raceways. Some raceways have removable covers, including wireways and some surface raceways.

Reactance. The opposition to the flow of current in an alternating current circuit that is due to the combination of capacitance and inductance. (*See also* Impedance)

Receptacle. Device installed at an outlet for the connection of a single plug. (*See also* Duplex receptacle)

Resistance (R). Measure, in ohms, of opposition to current flow due to the inherent electrical characteristics of a conductor, as distinguished from the effect of magnetic fields or the proximity of other conductors. In direct current, volts ÷ amperes = ohms resistance. (*See also* Impedance)

Romex. Trade name for Type NM nonmetallic-sheathed cable.

Series circuit. Connection of two or more devices or loads in tandem so that the current flowing through each also flows through all the others. Rarely used in residential and farm wiring.

Service entrance. That portion of the system that provides the interface between the distribution lines of the power-serving agency and the premises wiring, including the main circuit breaker or fused switch and the associated wiring.

Service equipment. Main fixed control and cutoff of the electrical supply consisting of circuit breaker(s) or switch(es) and fuses and their accessories and connected to the load end of service conductors in or on a building.

Shock. Accidental flow of electric current through the body of a human or animal.

Short circuit. An unintended, low-impedance, direct line-to-line connection. If not cleared immediately, these events can be extremely destructive.

Single-phase current. Current produced from a single alternating source; typical residential current provided by power suppliers, using two wires (for 120-volt only) or three wires (120/240-volt).

Source. Starting point of the circuit or system under consideration. For a wiring system the source is the power-serving agency's transformer; for a branch circuit it is the overcurrent device at the panelboard; for an added outlet it is the existing outlet from which the extension is made.

Subpanel. A panelboard supplied by a feeder connected to another panelboard;

the originating panelboard is usually, but not necessarily, the panelboard that is part of the service equipment. Subpanels are panelboards; the term "subpanel" is no longer used in the *NEC*.

Splice. Two or more wires connected together mechanically and electrically to provide good conductivity between them. (*See also* Pigtail)

Stray current. Undesirable current, usually through the earth, resulting either from multiple system grounds or from insulation failure. Especially hazardous to large farm animals.

Switch. Device for turning electrical equipment on or off by closing or opening the circuit to the equipment; designed for use at an amperage and voltage not higher than the limits marked on the equipment.

Tap. The point where a conductor is connected to another, usually larger, conductor mid-run to supply a load.

Three-phase current (power). Current produced from three separate alternating forces, each peaking at a different moment in time but in strict regularity, one after the other, that create naturally rotating fields in motors and typically used by factories and comparable establishments. Also common in establishments that have large concentrated loads, as in shopping areas or large apartment houses. Supplied through three ungrounded wires and usually a grounded wire.

Three-phase motor. Electric motor that operates only on three-phase current.

Three-way switches. Switches having three terminals for wires and used in pairs to control a single light from two locations.

Three-wire circuits. Circuit that is the equivalent of two 2-wire circuits sharing the grounded wire for the purpose of reducing material costs and voltage drop. Also called a multiwire circuit.

Three-wire cord. Cord containing a grounding wire in addition to the two circuit wires; the grounding wire is connected to the third prong of a three-prong plug that can be used only with a grounding-type (three-wire) receptacle. The purpose of the grounding wire is to reduce shock hazard.

Three-wire entrance (single-phase, 120/240-volt service). Three wires (two hot and one grounded) run from power supplier's transformer to building served, making available both 120-volt and 240-volt power.

Three-wire receptacle. Grounding-type receptacle having a third opening for the third (grounding) prong on a three-prong plug. Reduces shock hazard when used with correctly wired three-wire cord and three-prong plug sets.

Transformer. Device used to raise or lower ac voltage. Volts × amperes in the coil of the primary winding equal volts × amperes in the coil of the secondary winding, less the small current necessary to magnetize the laminated steel core.

Underwriters Laboratories Inc. (UL). The most widely known qualified testing laboratory that lists and labels electrical wiring materials and utilization equipment when they meet nationally recognized standards.

Volt (V).   Unit used to express the electromotive force (emf) that causes current to flow. Water analogy: pounds per square inch of water pressure.

Voltage drop.   Loss of voltage in a wire between the source and the end, caused by the resistance of the wire and the length of the run.

Watt (W).   The unit used to express electrical power. In direct current work, one volt × one ampere = one watt.

Wire.   Conductor, circular in cross section, through which electric current flows.

INDEX